Lecture Notes in Mathematics 1891

Editors:
J.-M. Morel, Cachan
F. Takens, Groningen
B. Teissier, Paris

Fondazione C.I.M.E. Firenze

C.I.M.E. means Centro Internazionale Matematico Estivo, that is, International Mathematical Summer Center. Conceived in the early fifties, it was born in 1954 and made welcome by the world mathematical community where it remains in good health and spirit. Many mathematicians from all over the world have been involved in a way or another in C.I.M.E.'s activities during the past years.

So they already know what the C.I.M.E. is all about. For the benefit of future potential users and co-operators the main purposes and the functioning of the Centre may be summarized as follows: every year, during the summer, Sessions (three or four as a rule) on different themes from pure and applied mathematics are offered by application to mathematicians from all countries. Each session is generally based on three or four main courses (24−30 hours over a period of 6-8 working days) held from specialists of international renown, plus a certain number of seminars.

A C.I.M.E. Session, therefore, is neither a Symposium, nor just a School, but maybe a blend of both. The aim is that of bringing to the attention of younger researchers the origins, later developments, and perspectives of some branch of live mathematics.

The topics of the courses are generally of international resonance and the participation of the courses cover the expertise of different countries and continents. Such combination, gave an excellent opportunity to young participants to be acquainted with the most advance research in the topics of the courses and the possibility of an interchange with the world famous specialists. The full immersion atmosphere of the courses and the daily exchange among participants are a first building brick in the edifice of international collaboration in mathematical research.

C.I.M.E. Director
Pietro ZECCA
Dipartimento di Energetica "S. Stecco"
Università di Firenze
Via S. Marta, 3
50139 Florence
Italy
e-mail: zecca@unifi.it

C.I.M.E. Secretary
Elvira MASCOLO
Dipartimento di Matematica
Università di Firenze
viale G.B. Morgagni 67/A
50134 Florence
Italy
e-mail: mascolo@math.unifi.it

For more information see CIME's homepage: http://www.cime.unifi.it

CIME's activity is supported by:

− Ministero degli Affari Esteri, Direzione Generale per la Promozione e la Cooperazione, Ufficio V
− Ministero dell'Istruzione, Università e Ricerca, Consiglio Nazionale delle Ricerche
− E.U. under the Training and Mobility of Researchers Programme

J.B. Friedlander · D.R. Heath-Brown
H. Iwaniec · J. Kaczorowski

Analytic Number Theory

Lectures given at the
C.I.M.E. Summer School
held in Cetraro, Italy,
July 11–18, 2002

Editors: A. Perelli, C. Viola

 Springer

Fondazione
C.I.M.E.

Authors and Editors

J.B. Friedlander
Department of Mathematics
University of Toronto
40 St George street
Toronto, ON M5S 2E4
Canada
e-mail: frdlndr@math.toronto.edu

D.R. Heath-Brown
Mathematical Institute
University of Oxford
24-29 St Giles
Oxford OX1 3LB
England
e-mail: rhb@maths.ox.ac.uk

H. Iwaniec
Department of Mathematics
Rutgers University
110 Frelinghuysen road
Piscataway, NJ 08854
USA
e-mail: iwaniec@math.rutgers.edu

J. Kaczorowski
Faculty of Mathematics and Computer
Science
Adam Mickiewicz University
ul. Umultowska 87
61-614 Poznan
Poland
e-mail: kjerzy@amu.edu.pl

Alberto Perelli
Dipartimento di Matematica
Università di Genova
Via Dodecaneso 35
16146 Genova
Italy
e-mail: perelli@dima.unige.it

Carlo Viola
Dipartimento di Matematica
Università di Pisa
Largo Pontecorvo 5
56127 Pisa
Italy
e-mail: viola@dm.unipi.it

Library of Congress Control Number: 2006930414

Mathematics Subject Classification (2000): 11D45, 11G35, 11M06, 11M20, 11M36, 11M41, 11N13, 11N32, 11N35, 14G05

ISSN print edition: 0075-8434
ISSN electronic edition: 1617-9692
ISBN-10 3-540-36363-7 Springer Berlin Heidelberg New York
ISBN-13 978-3-540-36363-7 Springer Berlin Heidelberg New York

DOI 10.1007/3-540-36363-7

Springer is a part of Springer Science+Business Media
springer.com
© Springer-Verlag Berlin Heidelberg 2006

Typesetting by the authors and SPi using a Springer LATEX package
Cover design: WMXDesign GmbH, Heidelberg

Printed on acid-free paper   SPIN: 11795704   41/SPi   5 4 3 2 1 0

Preface

The origins of analytic number theory, i.e. of the study of arithmetical problems by analytic methods, can be traced back to Euler's 1737 proof of the divergence of the series $\sum 1/p$ where p runs through all prime numbers, a simple, yet powerful, combination of arithmetic and analysis. One century later, during the years 1837-40, Dirichlet produced a major development in prime number theory by extending Euler's result to primes p in an arithmetic progression, $p \equiv a \pmod{q}$ for any coprime integers a and q. To this end Dirichlet introduced group characters χ and L-functions, and obtained a key result, the non-vanishing of $L(1, \chi)$, through his celebrated formula on the number of equivalence classes of binary quadratic forms with a given discriminant.

The study of the distribution of prime numbers was deeply transformed in 1859 by the appearance of the famous nine pages long paper by Riemann, *Über die Anzahl der Primzahlen unter einer gegebenen Grösse*, where the author introduced the revolutionary ideas of studying the zeta-function $\zeta(s) = \sum_1^\infty n^{-s}$ (and hence, implicitly, also the Dirichlet L-functions) as an analytic function of the complex variable s satisfying a suitable functional equation, and of relating the distribution of prime numbers with the distribution of zeros of $\zeta(s)$. Riemann considered it highly probable ("sehr wahrscheinlich") that the complex zeros of $\zeta(s)$ all have real part $\frac{1}{2}$. This still unproved statement is the celebrated Riemann Hypothesis, and the analogue for all Dirichlet L-functions is known as the Grand Riemann Hypothesis. Several crucial results were obtained in the following decades along the way opened by Riemann, in particular the Prime Number Theorem which had been conjectured by Legendre and Gauss and was proved in 1896 by Hadamard and de la Vallée Poussin independently.

During the twentieth century, research subjects and technical tools of analytic number theory had an astonishing evolution. Besides complex function theory and Fourier analysis, which are indispensable instruments in prime number theory since Riemann's 1859 paper, among the main tools and contributions to analytic number theory developed in the course of last century one should mention at least the circle method introduced by Hardy, Littlewood and Ramanujan in the 1920's, and later improved by Vinogradov and by Kloosterman, as an analytic technique for the study of diophantine equations and of additive problems over primes or over special integer sequences, the sieve methods of Brun and Selberg, subsequently developed by Bombieri, Iwaniec and others, the large sieve introduced by Linnik and substantially modified and improved by Bombieri, the estimations of exponential sums due to Weyl, van der Corput and Vinogradov, and the theory of modular forms and automorphic L-functions.

The great vitality of the current research in all these areas suggested our proposal for a C.I.M.E. session on analytic number theory, which was held at Cetraro (Cosenza, Italy) from July 11 to July 18, 2002. The session consisted of four six-hours courses given by Professors J. B. Friedlander (Toronto), D. R. Heath-Brown (Oxford), H. Iwaniec (Rutgers) and J. Kaczorowski (Poznań). The lectures were attended by fifty-nine participants from several countries, both graduate students and senior mathematicians. The expanded lecture notes of the four courses are presented in this volume.

The main aim of Friedlander's notes is to introduce the reader to the recent developments of sieve theory leading to prime-producing sieves. The first part of the paper contains an account of the classical sieve methods of Brun, Selberg, Bombieri and Iwaniec. The second part deals with the outstanding recent achievements of sieve theory, leading to an asymptotic formula for the number of primes in certain thin sequences, such as the values of two-variables polynomials of type $x^2 + y^4$ or $x^3 + 2y^3$. In particular, the author gives an overview of the proof of the asymptotic formula for the number of primes represented by the polynomial $x^2 + y^4$. Such an overview clearly shows the role of bilinear forms, a new basic ingredient in such sieves.

Heath-Brown's lectures deal with integer solutions to Diophantine equations of type $F(x_1, \ldots, x_n) = 0$ with absolutely irreducible polynomials $F \in \mathbb{Z}[x_1, \ldots, x_n]$. The main goal here is to count such solutions, and in particular to find bounds for the number of solutions in large regions of type $|x_i| \leqslant B$. The paper begins with several classical examples, with the relevant problems for curves, surfaces and higher dimensional varieties, and with a survey of many results and conjectures. The bulk of the paper deals with the proofs of the main theorems where several tools are employed, including results from algebraic geometry and from the geometry of numbers. In the final part, applications to power-free values of polynomials and to sums of powers are given.

The main focus of Iwaniec's paper is on the exceptional Dirichlet character. It is well known that exceptional characters and exceptional zeros play a relevant role in various applications of the L-functions. The paper begins with a survey of the classical material, presenting several applications to the class number problem and to the distribution of primes. Recent results are then outlined, dealing also with complex zeros on the critical line and with families of L-functions. The last section deals with Linnik's celebrated theorem on the least prime in an arithmetic progression, which uses many properties of the exceptional zero. However, here the point of view is rather different from Linnik's original approach. In fact, a new proof of Linnik's result based on sieve methods is given, with only a moderate use of L-functions.

Kaczorowski's lectures present a survey of the axiomatic class S of L-functions introduced by Selberg. Essentially, the main aim of the Selberg class theory is to prove that such an axiomatic class coincides with the class of automorphic L-functions. Although the theory is rich in interesting conjectures, the focus of these lecture notes is mainly on unconditional results. After a chapter on classical examples of L-functions and one on the basic theory, the notes present an account of the invariant theory for S. The core of the theory begins with chapter 4, where the necessary material on hypergeometric functions is collected. Such results are applied in the following chapters, thus obtaining information on the linear and non-linear twists which, in turn, yield a complete characterization of the degree 1 functions and the non-existence of functions with degree between 1 and 5/3.

We are pleased to express our warmest thanks to the authors for accepting our invitation to the C.I.M.E. session, and for agreeing to write the fine papers collected in this volume.

Alberto Perelli   Carlo Viola

Contents

Producing Prime Numbers via Sieve Methods

John B. Friedlander

Department of Mathematics, University of Toronto
40 St George street, Toronto, ON M5S 2E4, Canada
e-mail: frdlndr@math.toronto.edu

These notes represent an expanded version of the lectures on sieve methods which were delivered at the C.I.M.E. summer school in analytic number theory in Cetraro, Italy during the period July 11 to July 18, 2002. As such they are produced here in the same informal style and with the same goals as were those lectures.

The basic purpose for which the sieve was invented was the successful estimation of the number of primes in interesting integer sequences. Despite some intermittent doubts that this could ever be achieved, the objective has in recent years finally been reached in certain cases. One main goal of these lectures was to provide an introduction to these developments. Such an introduction would not have been appropriate to many in the target audience without some of the relevant background and a second objective was the provision during the first half of the lectures of a quick examination of the development of sieve methods during the past century and of the main ideas involved therein. As a result of these twin goals, the second half of the material is necessarily a little more technical than is the first part. It is hoped that these notes will provide a good starting point for graduate students interested in learning about sieve methods who will then go on to a more detailed study, for example [Gr, HR], and also for mathematicians who are not experts on the sieve but who want a speedy and relatively painless introduction to its workings. In both groups it is intended to develop a rough feeling for what the sieve is and for what it can and cannot do.

The sieve has over the years come to encompass an extensively developed body of work and the goals of these notes do not include any intention to give a treatment which is at all exhaustive, wherein one can see complete proofs, nor even to provide a reference from which one can quote precise statements of the main theorems. For those purposes the references provided are more than sufficient.

Acknowledgements

Over the past thirty years the author has had on many occasions the opportunity to discuss the topic of sieve methods with many colleagues, in particular with A. Selberg, E. Bombieri, and most frequently of all with H. Iwaniec. Indeed the current notes, together with the lecture notes [Iw5], form the starting points for a book on the subject which Iwaniec and I have begun to write. After years of extensive collaborations one cannot help but include thoughts which originated with the other person; indeed one cannot always remember which those were.

During the preparation of this work the author has received the generous support of the Canada Council for the Arts through a Killam Research Fellowship and also from the Natural Sciences and Engineering Research Council of Canada through Research Grant A5123.

1 "Classical" sieve methods

Eratosthenes

The sieve begins with Eratosthenes. We let x be a positive integer and

$$\mathcal{A} = \{n \leqslant x\},$$

the set of integers up to x. We are going to count the number of primes in this set.

For purposes of illustration let us choose $x = 30$. Thus we begin with the integers

1	2	3	4	5	6	7	8	9	10	11	12	13	14	15
16	17	18	19	20	21	22	23	24	25	26	27	28	29	30

and from these we are going to delete the ones that are composite, counting the number that remain. Our first step is to cross out those that are even, the multiples of two. This leaves with the following picture.

Turning to the next prime, three, we cross out all of its multiples. This leaves us with the following.

Note that there are some numbers, namely the multiples of six, which have been crossed out twice. If we are keeping a count of what has been left behind we should really add these back in once. Next we progress to the next prime number, five, and delete the multiples of that one. This gives us the following picture.

1   ②   3̸   ④   5̸   6̸   7   ⑧   9̸   ⑩   11   ⑫   13   ⑭   1̸5̸
⑯   17   ⑱   19   ⑳   2̸1̸   ㉒   23   ㉔   2̸5̸   ㉖   2̸7̸   ㉘   29   3̸0̸

Here again we find more numbers, the multiples of ten and of fifteen, that have been removed twice and so should be added back in once to rectify the count. But now we have even come to a number, thirty, which has been crossed out as a multiple of each of three primes. In this case, it has been crossed out three times (once each as a multiple of two, three and five), then added back in three times (once each as a multiple of six, ten and fifteen). Since thirty is composite we want to remove it precisely once so we have now to subtract it out one more time.

We are now ready to proceed to the multiples of the next prime, seven. However, before we do so it is a very good idea to notice that all of the remaining numbers on our list, apart from the integer one, are themselves prime numbers. This is a consequence of the fact that every composite positive integer must be divisible by some number (and hence some prime number) which is no larger than its square root. In our case all of the numbers are less than or equal to thirty and hence we only need to cross out multiples of primes $p \leqslant \sqrt{30}$ and five is the largest such prime. As a result we are ready to stop this procedure.

Let's think about what we have accomplished. On the one hand, totalling up the results of the count of our inclusion–exclusion, we began (in the case $x = 30$) with $[x]$ integers, for each prime $p \leqslant \sqrt{x}$ we subtracted out $[x/p]$ multiples of p, then for each pair of distinct primes $p_1 < p_2 \leqslant \sqrt{x}$ we added back in the $[x/p_1p_2]$ multiples of p_1p_2, and so on. In all, we are left with the final count

$$[x] - \sum_{p \leqslant \sqrt{x}} \left[\frac{x}{p}\right] + \sum\sum_{p_1 < p_2 \leqslant \sqrt{x}} \left[\frac{x}{p_1 p_2}\right] - \sum\sum\sum_{p_1 < p_2 < p_3 \leqslant \sqrt{x}} \left[\frac{x}{p_1 p_2 p_3}\right] + - \cdots$$

On the other hand, this was after all just the count for the number of integers not crossed out and these integers are just the primes less than or equal to x, other than those which are less than or equal to $\sqrt{x}$, together with the integer one.

Equating the two we obtain the

Legendre Formula

$$\pi(x) - \pi(\sqrt{x}) + 1 = \sum_{\substack{d \\ p|d \Rightarrow p \leqslant \sqrt{x}}} \mu(d)\left[\frac{x}{d}\right].$$

Here, as usual, $\pi(x)$ denotes the prime counting function

$$\pi(x) = \sum_{p \leqslant x} 1,$$

and, throughout, the letter p will always be a prime. As usual, the Möbius function $\mu(d)$ is $(-1)^\nu$ when d is the product of $\nu \geqslant 0$ distinct primes and is zero if d has a repeated prime factor. This function provides a concise way of expressing the right hand side of the formula.

It will turn out that $\pi(x)$ is considerably larger than $\sqrt{x}$, hence (since trivially $\pi(\sqrt{x}) \leqslant \sqrt{x}$) the left side of the Legendre formula is approximately $\pi(x)$. In order to estimate $\pi(x)$ we thus want to develop the right side.

The obvious starting point for an estimation of the right hand side is the replacement everywhere of the awkward function $[t]$, the integral part of t, by the simpler function t. This makes an error of $\{t\}$, the fractional part. More precisely, we have

$$\text{right side} = x \sum_d \frac{\mu(d)}{d} + E = x \prod_{p \leqslant \sqrt{x}} \left(1 - \frac{1}{p}\right) + E$$

where the error term E is

$$E = -\sum_d \mu(d)\left\{\frac{x}{d}\right\}.$$

At first glance, the best we can expect to do is to use the trivial bound $\{t\} < 1$ which leads us to bound the error term by

$$|E| \leqslant \sum_d 1 = 2^{\pi(\sqrt{x})},$$

which is absolutely enormous, much larger even than the number of integers $[x]$ that we started with. Of course, we have been particularly stupid here, for example, sieving out multiples of d even for certain integers d exceeding x, so the above bound can certainly be improved somewhat. Unfortunately however, E is genuinely large. In fact, using old ideas due to Chebyshev and to Mertens, one knows that

$$\prod_{p \leqslant \sqrt{x}} \left(1 - \frac{1}{p}\right) \sim \frac{e^{-\gamma}}{\log \sqrt{x}}$$

so what we have been expecting to be our main term is actually wrong. Since, by the prime number theorem,

$$\pi(x) \sim \frac{x}{\log x},$$

we see that the quantity E we have been referring to as the error term has the same order of magnitude as the main term.

Brun

The sieve of Eratosthenes lay in such a state, virtually untouched for almost two thousand years. The modern subject of sieve methods really begins with Viggo Brun. Although he later developed significant refinements to what we shall describe here, Brun's first attempts to make the error term more manageable were based on the following quite simple ideas.

Although one cannot greatly improve the trivial bound in the error term for each individual d on the right side, one can try to cut down on the number of terms in the sum. One way to do this is to cut the process off earlier, sifting out multiples of primes only up to some chosen z which is smaller than $\sqrt{x}$. Moreover, re-examining the inclusion–exclusion procedure and truncating this, we see that, if we truncate after d with a specified even number of prime factors, say $\nu(d) = 2r$, we get an upper bound, while if we truncate after an odd number $\nu(d) = 2r + 1$, we get a lower bound.

Although not an asymptotic formula, such bounds can be valuable. For example, an upper bound will, a fortiori, provide an upper bound for $\pi(x) - \pi(z)$ and hence (when combined with the trivial bound $\pi(z) \leqslant z$) an upper bound for $\pi(x)$. A positive lower bound will demonstrate the existence of integers without any small prime factors, and hence with few prime factors (the latter are referred to as "almost-primes"). Thus for example, an integer $n \leqslant x$ having no prime factor $p \leqslant x^{1/4}$ can have at most three prime factors.

Some Generality

So far we are in the rather depressing position that we have a method which fails to give us good estimates for the number $\pi(x)$ of primes up to x, but even worse, the only reason we even know that it is doomed to fail is because other techniques, from analytic number theory, succeed (to prove the prime number theorem), thereby telling us so. What then is the value of the sieve is that it can be generalized to give some information in cases where the analytic machinery is lacking. Therefore, to consider the situation more generally is not merely worthwhile; it is the sieve's only raison d'être.

We consider a finite sequence of non-negative reals

$$\mathcal{A} = (a_n), \quad n \leqslant x,$$

and a set $\mathcal{P}$ of primes. It is convenient to denote

$$P(z) = \prod_{\substack{p \in \mathcal{P} \\ p < z}} p.$$

Our goal is to estimate the "sifting function"

$$S(\mathcal{A}, z) = \sum_{\substack{n \leqslant x \\ (n, P(z)) = 1}} a_n.$$

We proceed just as in our original example, but phrased in slightly different fashion. We need the basic property of the Möbius function

$$\sum_{d \mid n} \mu(d) = \begin{cases} 1 & \text{if } n = 1, \\ 0 & \text{if } n > 1. \end{cases}$$

We also use the simple fact from elementary number theory that $\delta \mid a$, $\delta \mid b \iff \delta \mid (a, b)$, that is, the set of common divisors of two positive integers is just the same as the set of divisors of their greatest common divisor.

Inserting these two facts and then interchanging the order of summation we obtain

$$S(\mathcal{A}, z) = \sum_n a_n \sum_{d \mid (n, P(z))} \mu(d) = \sum_n a_n \sum_{\substack{d \mid n \\ d \mid P(z)}} \mu(d)$$

$$= \sum_{d \mid P(z)} \mu(d) \sum_{n \equiv 0 \, (\mathrm{mod}\, d)} a_n = \sum_{d \mid P(z)} \mu(d) A_d(x),$$

say. This is just (a more general version of) the Legendre formula and here as before we need information about the sums

$$A_d(x) = \sum_{\substack{n \leqslant x \\ n \equiv 0 \, (d)}} a_n$$

which give the mass of the subsequence running over multiples of d, that is $\mathcal{A}_d = (a_{md})$, $m \leqslant x/d$, and which in our beginning example was $[x/d]$. Specifically, we need a useful approximation formula. We assume we can write this in the form

$$(*) \qquad\qquad A_d(x) = A(x) g(d) + r_d(x),$$

where

$$A(x) = A_1(x) = \sum_{n \leqslant x} a_n$$

is the total mass of our sequence, where $g(d)$ is a "nice" function (equal to $1/d$ in our example) and $r_d(x)$ is a "remainder" which is small, at least on average over d (this was $-\{x/d\}$ in our example). Inserting our approximation formula $(*)$ the sifting function becomes

$$S(\mathcal{A}, z) = A(x) \sum_{d|P(z)} \mu(d)g(d) + \sum_{d|P(z)} \mu(d)r_d(x)$$

which is basic to all that follows. The function $g(d)$ behaves like a probability in a number of respects, describing approximately the fraction of the total mass coming from multiples of d. (It is useful to keep in mind $g(d) = 1/d$ as the prototype for such a function.) Hence, we shall assume $g(1) = 1$ and that, for each $d > 1$, we have $0 \leqslant g(d) < 1$. If for some $d > 1$ we had $g(d) = 1$ virtually everything would be a multiple of d and there would not be much point in looking for primes. We also assume that g is a multiplicative function, that is whenever $(d_1, d_2) = 1$ we have

$$g(d_1 d_2) = g(d_1)g(d_2).$$

The essence of this is that we are assuming that divisibility by two relatively prime integers are independent events. In practice this is true only to a rather limited extent and this fact is in large measure responsible for the failure of the method to do better.

Some Examples

We consider some examples. In many of the most basic examples the sequence $\mathcal{A}$ is just the characteristic function of an interesting set of integers. In such a case we shall sometimes abuse notation by failing to distinguish between the function and the set on which it is supported.

Example 1 We begin by repeating once again our original example. Thus, we have

$$\mathcal{A} = \{m \mid m \leqslant x\}, \qquad \mathcal{P} = \{\text{all primes}\},$$

$$A_d(x) = \left[\frac{x}{d}\right] = \frac{x}{d} - \left\{\frac{x}{d}\right\},$$

$$g(d) = \frac{1}{d}, \qquad r_d(x) = -\left\{\frac{x}{d}\right\}.$$

Example 2 Now for something a little different, consider

$$\mathcal{A} = \{m^2 + 1 \leqslant x\}, \qquad \mathcal{P} = \{p, \ p \not\equiv 3 \ (\text{mod} \, 4)\},$$

$$g(p) = \begin{cases} 2/p & p \equiv 1 \ (\text{mod} \, 4) \\ 1/2 & p = 2, \end{cases} \qquad |r_d| \leqslant 2^{\nu(d)},$$

this last estimate following from the bound $|r_p| \leqslant 2$ and the Chinese Remainder Theorem. Here, there is no need to sieve by the primes congruent to three modulo four since none of the integers in our set is divisible by any such prime (although we could, equivalently, sieve by the set of all primes and simply set $g(p) = 0$ for these additional primes). In this example if we were able to get a positive lower bound for $S(\mathcal{A}, \sqrt{x}\,)$ we would be producing primes of the form $m^2 + 1$. It is a famous problem to show that there are infinitely many such primes.

Example 3 For another famous conjecture, we consider the following example.

$$\mathcal{A} = \{m(m+2) \leqslant x\}, \qquad \mathcal{P} = \{\text{all primes}\},$$

$$g(p) = \begin{cases} 2/p & p \text{ odd} \\ 1/2 & p = 2, \end{cases} \qquad |r_d| \leqslant 2.$$

Here, if we could give a positive lower bound for $S(\mathcal{A}, x^{1/4})$ we would be producing integers $m(m+2)$ where both factors are prime and differ by two. The "twin prime conjecture" predicts that there are infinitely many such pairs of primes.

Example 4 There is an alternative appraoch via the sieve to attack this last conjecture. As our fourth example we consider the following sequence.

$$\mathcal{A} = \{p - 2 \leqslant x\}, \qquad \mathcal{P} = \{\text{odd primes}\},$$

$$A_d(x) = \pi(x; d, 2),$$

$$g(p) = \frac{1}{p-1}, \qquad g(d) = \frac{1}{\varphi(d)},$$

where $\pi(x; d, a)$ is the number of primes up to x which are congruent to a modulo d and where $\varphi(d)$, the Euler function, counts the number of units in the ring of residue classes modulo d. This example offers some advantages over the previous one for studying the twin prime problem and at this point in time it gives stronger results, although this was not always the case. Most significantly, we are starting from the beginning with the knowledge that one of our two numbers p, $p-2$ is a prime. On the other hand, the remainder term is more complicated, namely $r_d(x) = \pi(x; d, 2) - \pi(x)/\varphi(d)$, and it is much more difficult to bound it successfully. In the current state of knowledge, a reasonably good bound can only be given on average over d; the most famous bound of this type being the celebrated Bombieri–Vinogradov theorem [Bo1]. Once again, if we could be successful in giving a positive lower bound, this time for $S(\mathcal{A}, \sqrt{x}\,)$, then we would produce twin primes.

It is possible to give many more examples wherein well-known problems concerning primes, for instance the Goldbach conjecture, can be phrased so as to follow from sufficiently strong sieve-theoretic estimates. Phrasing them this way is however by far the easier part of the problem.

Upper and Lower Bounds

Let us return to the general version of the Legendre formula, namely

$$S(\mathcal{A}, z) = \sum_{d|P(z)} \mu(d) A_d(x)$$

$$= A(x) \sum_{d|P(z)} \mu(d) g(d) + \sum_{d|P(z)} \mu(d) r_d(x).$$

Recall that this was based on the basic property:

$$\sum_{d|n} \mu(d) = \begin{cases} 1 & \text{if } n = 1 \\ 0 & \text{if } n > 1 \end{cases}$$

$= \Theta_0(n)$, say.

Now suppose that we replace the Möbius function $\mu(d)$ by one of two sequences λ_d^+, λ_d^-, which we shall refer to as the "sifting weights", these having the properties

$$\Theta^+(n) \doteq \sum_{d|n} \lambda_d^+ \begin{cases} = 1 & \text{if } n = 1 \\ \geq 0 & \text{if } n > 1, \end{cases}$$

or alternatively

$$\Theta^-(n) \doteq \sum_{d|n} \lambda_d^- \begin{cases} = 1 & \text{if } n = 1 \\ \leq 0 & \text{if } n > 1. \end{cases}$$

Repeating the same manipulation as before we obtain upper and lower bounds

$$S^-(\mathcal{A}, z) \leq S(\mathcal{A}, z) \leq S^+(\mathcal{A}, z)$$

where

$$S^+(\mathcal{A}, z) = \sum_n a_n \Theta^+(n) = \sum_d \lambda_d^+ A_d(x)$$

$$= A(x) \sum_d \lambda_d^+ g(d) + \sum_d \lambda_d^+ r_d(x),$$

and similarly

$$S^-(\mathcal{A}, z) = A(x) \sum_d \lambda_d^- g(d) + \sum_d \lambda_d^- r_d(x).$$

Recall the problem we had before with the remainder term

$$\sum_{d|P(z)} \mu(d)r_d(x).$$

Now we can force this term to be small by insisting that $\lambda_d^{\pm}$ vanish beyond a certain point, say for $d > D$. This is not something we could do with the Möbius function which was a unique function given to us by nature. Now we have a whole family of functions, conceivably quite a lot of them, and perhaps even after truncating them in this fashion there will still remain a number of reasonable choices.

Problem *How do we choose the sifting weights $\{\lambda_d\}$?*

We want to choose a sequence λ_d^+, $d \leqslant D$, so that S^+ is minimal, or at least fairly small (and a sequence λ_d^- with the corresponding properties for S^-). This is a very complicated problem so we make some simplifying assumptions. Although each of $S^{\pm}$ has two sums in it we are going to attempt to choose $\lambda_d^{\pm}$ so that the sum $\sum_d \lambda_d^{\pm} g(d)$, and hence the main term $A(x) \sum_d \lambda_d^{\pm} g(d)$, is satisfactory and simply hope that after this choice has been specified the remainder term will also turn out to be acceptable. Once we make this simplifying assumption the individual properties of the sequence $\mathcal{A}$ are removed from consideration in the choice of weights. Thus, the choice of weights $\lambda_d^{\pm}$, whatever that may be, should be the same for all sequences $\mathcal{A}$ which give rise to the same function g.

The first successful sieve, Brun's "pure" sieve, made choices of the type

$$\lambda_d^+ = \begin{cases} \mu(d) & \nu(d) \leqslant 2r, \\ 0 & \text{else}, \end{cases} \qquad \lambda_d^- = \begin{cases} \mu(d) & \nu(d) \leqslant 2r+1, \\ 0 & \text{else}, \end{cases}$$

for suitable r. Subsequently Brun discovered some considerably more refined sieve weights which are however much more complicated to describe. Although this first sieve does not give results as strong as those later ones it was sufficient for Brun [Br] to prove the first striking application of the theory:

Theorem (Brun) *The sum of the reciprocals of the twin primes is convergent, that is*

$$\sum_{p \text{ twin}} \frac{1}{p} < \infty.$$

Selberg's Upper Bound

There are many other possibilities for the choice of sieve weights. One of the best known, due to Selberg [Se1], is an upper bound sieve.

Consider any set $\rho = \{\rho_d, \ d|P(z)\}$ of real numbers satisfying $\rho_1 = 1$ and $\rho_d = 0$ for all $d > \sqrt{D}$. Then, for any $n|P(z)$ we certainly have

$$\left(\sum_{d|n} \rho_d \right)^2 \begin{cases} = 1 & \text{if } n = 1 \\ \geqslant 0 & \text{if } n > 1, \end{cases}$$

precisely the inequalities we required for a function $\Theta^+(n)$. At first glance this might not seem to be of the right shape, due to the square, but indeed it is of the form

$$S^+(\mathcal{A}, z) = \sum_{d|P(z)} \lambda_d^+ A_d(x)$$

with the coefficients given by

$$\lambda_d^+ = \sum_{\substack{d_1|P(z), \, d_2|P(z) \\ [d_1, d_2] = d}} \rho_{d_1} \rho_{d_2},$$

where $[d_1, d_2]$ denotes the least common multiple. Note that $\lambda_d^+ = 0$ for $d > D$.

Manipulating the above expression in the by now familiar fashion, we obtain the inequality

$$S(\mathcal{A}, z) \leqslant \sum_n a_n \left(\sum_{\substack{d|n \\ d|P(z)}} \rho_d \right)^2 = \sum_{\substack{d_1|P(z) \\ d_2|P(z)}} \sum \rho_{d_1} \rho_{d_2} A_{[d_1, d_2]}(x).$$

Now, substituting in the approximation formula for $A_{[d_1, d_2]}(x)$, and ignoring the remainder term as before, we see that the problem of choosing the weights λ_d^+, or equivalently the coefficients ρ_d, is just that of minimizing, for given g, the quadratic form

$$\sum_{\substack{d_1|P(z) \\ d_2|P(z)}} \sum \rho_{d_1} \rho_{d_2} \, g([d_1, d_2])$$

in the variables $\{\rho_d, \ d \leqslant \sqrt{D} \}$.

This extremal problem can be solved in a number of ways, for example using Lagrange multipliers as was done by Selberg. Although the result gives the best possible choice of upper bound weights only within the subfamily of weights having this special shape, it turned out that the solution provided gave, in many cases, dramatic improvements on what had been previously known, and even some results which have since been shown to be close to best possible.

Buchstab Iteration

Partly because of the simplicity and success of the above construction, but perhaps also partly due to the enormous influence of its inventor, the Selberg sieve received a great deal of attention, so much so that, in the works of a number of authors, the terms "sieve" and "Selberg sieve" came for a while to be used almost interchangeably (never of course by Selberg himself).

Nevertheless, despite its success the Selberg sieve had a drawback not shared by the other leading methods, before and since. In making essential use of the non-negativity of the squares of real numbers, it introduced an asymmetry between the upper and lower bound methods; there was no corresponding lower bound sieve which was quite as simple nor as successful as was the upper bound.

In view of this it became even more important that a method introduced by Buchstab [Bu], which had in fact predated Selberg's work, allowed one to deduce lower bounds from upper bounds (and vice-versa).

The idea behind this technique may be described as follows. Suppose that we are given a sequence $\mathcal{A}$ and two parameters $z_1 < z_2$. We consider the difference $S(\mathcal{A}, z_1) - S(\mathcal{A}, z_2)$ which counts the contribution of those elements which survive the sieve up to z_1 but do not survive it up to z_2. We group these terms in accordance with the smallest prime p which removes it. Such an element is of the form a_n where n is divisible by p but by no smaller prime in $\mathcal{P}$. Hence we deduce the Buchstab identity

$$S(\mathcal{A}, z_2) = S(\mathcal{A}, z_1) - \sum_{\substack{z_1 \leqslant p < z_2 \\ p \in \mathcal{P}}} S(\mathcal{A}_p, p).$$

Here, if we input an upper bound for each term in the above sum over p together with a lower bound for $S(\mathcal{A}, z_1)$ we obtain a lower bound for $S(\mathcal{A}, z_2)$. In practice, if z_1 is chosen to be relatively small then one can estimate $S(\mathcal{A}, z_1)$ fairly accurately, both above and below, and in consequence we may think of the first term on the right hand side as being a known quantity.

Similarly, a lower bound for each term in the sum over p combined with an upper bound for $S(\mathcal{A}, z_1)$ gives an upper bound for $S(\mathcal{A}, z_2)$. This is of somewhat less interest in connection with the Selberg sieve but is equally important in many other circumstances.

Of course, once such a procedure proves to be successful it seems natural that one should attempt to iterate it and, under the proper circumstances, further improvements do take place. Using these ideas and taking as starting points such results as the upper bound of Selberg, the trivial lower bound of zero, and the bounds, both lower and upper, of Brun in the range of small z where these are quite accurate, very good choices for the sieve weights were given during the decade of the 1960's. The first of these is due to Jurkat and Richert [JR] and then a different set of weights was found, independently by Iwaniec [Iw1, Iw3] and (as reported in Selberg [Se3]) by Rosser.

To describe the flavour of these results we consider the following diagram.

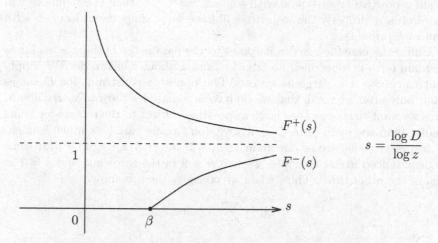

Here, the probabilistically expected answer for our sifting function is

$$S(\mathcal{A}, z) \approx A(x) \prod_{p \mid P(z)} \left(1 - g(p)\right)$$

which corresponds to the horizontal line at height one so that the difference between $F^+(s)$ and this line represents the deviation above and below this expectation which we are forced to tolerate in the main terms of our bounds $S^\pm(\mathcal{A}, z)$. We obtain

$$\sum_d \lambda_d^+ g(d) = F^+(s) \prod_{p \mid P(z)} \left(1 - g(p)\right) + \text{small},$$

$$\sum_d \lambda_d^- g(d) = F^-(s) \prod_{p \mid P(z)} \left(1 - g(p)\right) + \text{small}.$$

It is evident from the diagram that the results are better, that is closer to the expected value, when the variable s is large. This suggests that we should like to take D as large as possible and z as small as possible. There is however a countervailing force pushing z in the opposite direction; the larger we can take z and still get a positive lower bound (that is with $s > \beta$), the stronger the qualitative information we have about the size (and hence the number) of prime factors that we can guarantee some members of the sequence will possess.

The main point is the following. As long as we can, as happens in virtually every case of interest, choose each of D and z to be some fixed powers of x then we obtain an upper bound of the right order of magnitude for the contribution $\sum_{p \leqslant x} a_p$ from the primes in our sequence $\mathcal{A}$ and, on the other hand, the existence therein, for some fixed k, of integers having at most k prime factors (with a better value of k the larger we can choose z). For example, if we can obtain a positive lower bound with some $z > x^{1/(k+1)}$ then the sequence will contain (more properly the sequence will have some support on) integers with at most k prime factors.

Unlike the case for z, when it comes to the parameter D there is, as far as the main term is concerned, no mixed emotion about what to do. We simply want to choose it as large as we can. The opposing constraint for D comes about only after we recall that we do have a remainder term to worry about. Thus we want to choose D as large as possible subject to the remainder being smaller than the main term. Usually, we don't really care how much smaller.

In practice the size of our main term will be $\approx A(x)(\log x)^{-\kappa}$, for some small κ; indeed in our examples we had $\kappa = 2$ in the third one and $\kappa = 1$ in each of the other three. Thus, when we consider the remainder

$$R(D) = \sum_{d \leqslant D} \lambda_d \, r_d(x),$$

a reasonable goal is the bound

$$R(D) \ll A(x)(\log x)^{-A}$$

which we ask to hold for every $A > 0$. Usually, in cases where we can prove anything at all, we can prove this much.

For almost all of the basic sieve weights one would consider it turns out that we have $|\lambda_d| \leqslant 1$. (An exception we shall ignore is provided by the Selberg weights which still satisfy a bound almost that good.) As a result the remainder term satisfies the so-called "trivial bound"

$$|R(D)| \leqslant \sum_{d \leqslant D} |\lambda_d \, r_d(x)| \leqslant \sum_{d \leqslant D} |r_d(x)|.$$

When using this bound it is unreasonable to expect ever to achieve a successful outcome such as $R(D) \ll A(x)(\log x)^{-A}$ with any value of D exceeding $A(x)$, since this would imply a great many of the individual terms $r_d(x)$ are unreasonably small. On the other hand, as we shall see in the next chapter, it is sometimes possible, in cases where $A(x)$ is small compared to x, as in the second and third examples, that we can do better by not using the above trivial bound. Nevertheless, essential problems remain.

Parity Problem

The most important stumbling block in sieve theory during the past few decades has undoubtedly been the parity problem. This phenomenon was first observed by Selberg [Se2] who gave a number of interesting counterexamples which set limitations to what one could hope to accomplish with classical sieve methods. We mention only the simplest of these examples.

Consider the sequence

$$\mathcal{A} = \{m \leqslant x, \ \nu(m) \text{ even}\},$$

where, as we recall, $\nu(m) = \sum_{p|m} 1$. In this case $\mathcal{A}$ has no primes at all! On the other hand, this sequence has very regular properties of distribution in arithmetic progressions and it can be shown that $\mathcal{A}$ satisfies all of the classical sieve axioms including the necessary remainder term bound with D essentially as large as one could reasonably hope.

To see this problem most clearly we shall describe an alternative formulation of the sieve due to Bombieri.

Bombieri's Sieve

Up to now we have been studying the sum $\sum_{p \leqslant x} a_p$. More honestly, we had been hoping to study this sum but actually have spent most of our time on the more modest goal

$$\sum_{\substack{n \leqslant x \\ (n, P(z)) = 1}} a_n,$$

which would be essentially the same if only we were able to choose $z > \sqrt{x}$.

Now, instead we shall study the sum

$$S - \sum_{n \leqslant x} u_n \Lambda(n)$$

where $\Lambda(n)$ is the von Mangoldt function, introduced by Chebyshev [Tc],

$$\Lambda(n) = \begin{cases} \log p & n = p^r, \ r \geqslant 1 \\ 0 & \text{else,} \end{cases}$$

which is almost the same as studying the sum $\sum_{p \leqslant x} a_p \log p$. This is a rather small change but it works out a little better in some respects. Perhaps that is not a surprise when we recall how in using the analytic methods to study primes the von Mangoldt function turns out to be the natural weight.

Just as with our earlier functions $\Theta^{\pm}(n)$ we write $\Lambda(n)$ as a sum over the divisors of n:

$$\Lambda(n) = \sum_{d|n} \lambda_d,$$

but here

$$\lambda_d = -\mu(d) \log d,$$

so that in this case the sieve weights λ_d are rather natural functions, not the mysterious ones encountered earlier. Interchanging the order of summation as we have done several times before we obtain the same formula

$$S = \sum_{d \leqslant x} \lambda_d \, A_d(x)$$

$$= A(x) \sum_{d \leqslant x} \lambda_d \, g(d) + \sum_{d \leqslant x} \lambda_d \, r_d(x).$$

One advantage of our new formulation is that, in contrast to what we initially saw with the sieve of Eratosthenes, the main term now gives the right answer for the expected result! In fact, even if we have to restrict our consideration to small values of d, say $d < x^\varepsilon$, the main term still gives the expected result. Not to get too excited however. We still have the same very serious problem with the remainder term.

That problem will prevent us from detecting primes. How then do we adopt a fallback position? Before, in our earlier formulation, we merely truncated at a smaller level z which led us to study the distribution of almost-primes. A very natural way to proceed in this case is to introduce the generalized von Mangoldt functions. These are the Dirichlet convolutions

$$\Lambda_k(n) = \left(\mu * \log^k \right)(n) = \sum_{d|n} \mu(d) \left(\log \frac{n}{d} \right)^k.$$

That these functions generalize the von Mangoldt function is clear since $\Lambda_1 = \Lambda$. That they represent an analogue to our earlier notion of almost-primes follows since it turns out that the support of Λ_k is on integers having at most k distinct prime factors. This and a number of other nice properties follow from the recurrence formula

$$\Lambda_{k+1} = \Lambda_k L + \Lambda_k * \Lambda,$$

which holds for every $k \geqslant 1$ and is easily proved by induction on k. Here L denotes the arithmetic function $L(n) = \log n$.

We have

$$0 \leqslant \Lambda_k(n) \leqslant (\log n)^k,$$

the first inequality following from induction on k and using the recurrence, while the second one follows from the first since, by Möbius inversion, $\sum_{d|n} \Lambda_k(d) = (\log n)^k$.

For $k = 2$ we have

$$\Lambda_2(n) = \Lambda(n) \log n + \sum_{d|n} \Lambda(d) \Lambda\left(\frac{n}{d} \right),$$

a function which came to notice when Selberg gave an elementary proof of the asymptotic formula $\sum_{n \leqslant x} \Lambda_2(n) \sim 2x \log x$ and this idea in turn played a fundamental role in all of the first few elementary proofs of the prime number theorem.

From this formula it is evident that half of the mass of the sum $\sum_{n\leqslant x}\Lambda_2(n)$ comes from integers having an even number of prime factors and half from those with an odd number. It turns out the same dichotomy holds as well for the sum $\sum_{n\leqslant x}\Lambda_k(n)$ for each $k\geqslant 3$.

We now wish to study the general sum

$$S_k(x) = \sum_{n\leqslant x} a_n\Lambda_k(n).$$

It is expected that the above dichotomy holds here as well, at least for very general sequences, and this is indeed the essence of the parity problem.

We make a number of assumptions about the "niceness" of the function g occurring in the main term of the approximation formula for our sequence. In particular we require an assumption which says that $g(p) = 1$ on average; this may be phrased for example as requiring that

$$\sum_{p\leqslant x} g(p) \sim \log\log x.$$

We also make a very strong assumption about the "level of distribution" D for which the remainder term satisfies an adequate bound:

(R) $$\sum_{d\leqslant D} |r_d| \ll A(x)(\log x)^{-B},$$

for all $B > 0$.

Theorem (Bombieri [Bo2]**)** *Fix an integer $k\geqslant 2$. Assume that for every $\varepsilon > 0$ the bound* (R) *holds with $D = x^{1-\varepsilon}$, where the implied constant may depend on ε as well as B. Then*

$$\sum_{n\leqslant x} a_n\Lambda_k(n) \sim k\,H A(x)(\log x)^{k-1}$$

where $H = \prod_p \left(1 - g(p)\right)\left(1 - \dfrac{1}{p}\right)^{-1}.$

Note that this theorem is, by the Selberg counter-example, not true for $k = 1$. In case, for a given sequence $\mathcal{A}$, the corresponding asymptotic does hold for $k = 1$ but with a multiplicative factor α, that is

$$\sum_{n\leqslant x} a_n\Lambda(n) \sim \alpha\,H A(x),$$

then Bombieri showed that $0\leqslant \alpha \leqslant 2$, and that, for each $k\geqslant 2$, the weight in the asymptotic formula for the sum $\sum_{n\leqslant x} a_n\Lambda_k(n)$ coming from integers with an odd number of prime factors is α times the expected amount, so then of course the weight coming from integers with an even number of prime factors is $2 - \alpha$ times the expected amount. Examples show that every α in the range $0 \leqslant \alpha \leqslant 2$ can occur.

This theorem of Bombieri is also optimal in that it becomes false as soon as the range in the assumption (R) is relaxed to $D = x^{\vartheta}$ for some fixed $\vartheta < 1$. This is shown by the counterexamples provided in the very recent work [Fd] of Ford. These examples become increasingly delicate as ϑ approaches 1.

2 Sieves with cancellation

Basic Sieve Problem

We begin by quickly recapitulating our problem. We are given a finite sequence
$$\mathcal{A} = (a_n), \quad n \leqslant x,$$
of non-negative real numbers and a set $\mathcal{P}$ of primes. We denote
$$P(z) = \prod_{\substack{p \in \mathcal{P} \\ p < z}} p.$$

Our goal is to estimate the sifting function
$$S(\mathcal{A}, z) = \sum_{\substack{n \leqslant x \\ (n, P(z)) = 1}} a_n.$$

We define the "congruence sums"
$$A_d(x) \doteq \sum_{\substack{n \leqslant x \\ n \equiv 0 \,(\mathrm{mod}\, d)}} a_n,$$

which register the weight contributed by integer multiples of d. We assume the congruence sums have an approximation formula

$$(*) \qquad A_d(x) = A(x)g(d) + r_d(x)$$

where g is a nice function and r_d is not too large, at least on average over d.

We introduce two sequences $\lambda_d^{\pm}$ of real numbers supported on positive integers $d \leqslant D$, having $\lambda_1^{\pm} = 1$, and satisfying, for all $n > 1$, the conditions

$$\sum_{d|n} \lambda_d^+ \geqslant 0, \qquad \sum_{d|n} \lambda_d^- \leqslant 0.$$

We compute

$$S(\mathcal{A}, z) = \sum_{n \leqslant x} a_n \sum_{d | (n, P(z))} \mu(d) \leqslant \sum_{n \leqslant x} a_n \sum_{d | (n, P(z))} \lambda_d^+$$

$$= \sum_{n \leqslant x} a_n \sum_{\substack{d | n \\ d | P(z)}} \lambda_d^+ = \sum_{d | P(z)} \lambda_d^+ A_d(x)$$

and similarly

$$S(\mathcal{A}, z) \geqslant \sum_{d | P(z)} \lambda_d^- A_d(x).$$

Substituting in this the approximation formula (∗) for $A_d(x)$ we deduce the upper and lower bounds

$$S^-(\mathcal{A}, z) \leqslant S(\mathcal{A}, z) \leqslant S^+(\mathcal{A}, z)$$

where

$$S^{\pm}(\mathcal{A}, z) = A(x) \sum_{d} \lambda_d^{\pm} g(d) + \sum_{d} \lambda_d^{\pm} r_d(x)$$

$$= \text{main term} + \text{remainder}.$$

The Main Term

Recall that $g(d)$ is assumed to be a multiplicative function, much like a probability, hence in particular satisfying $0 \leqslant g(d) < 1$. In practice we shall also have

$$\sum_{p \leqslant x} g(p) \log p \sim \kappa \log x$$

for some constant $\kappa \geqslant 0$ which we call the "sifting density". We think of $g(d) = 1/d$ as our prototypical example; in this case the above assumption reduces to the formula

$$\sum_{p \leqslant x} \frac{\log p}{p} \sim x$$

which is an old theorem due to Chebyshev as is the equivalent asymptotic formula $\sum_{p \leqslant x} 1/p \sim \log \log x$ mentioned earlier. We think of κ as representing the average number of residue classes sifted by a typical prime p.

The main terms are described by the sums

$$\sum_{d} \lambda_d^+ g(d) = F^+(s) \prod_{p | P(z)} (1 - g(p)) + \text{small},$$

$$\sum_{d} \lambda_d^- g(d) = F^-(s) \prod_{p | P(z)} (1 - g(p)) + \text{small},$$

where the functions $F^{\pm}$ are given by the following diagram.

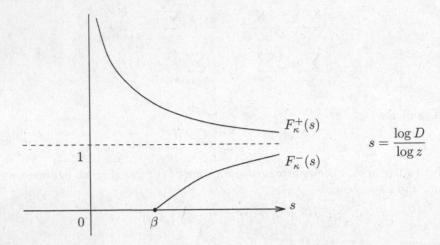

Here $F^+ = F^+_{\kappa}$, $F^- = F^-_{\kappa}$ may actually depend on κ.

For the specific values $\kappa = 1$, $\kappa \leqslant 1/2$ one knows sieves which give best possible results in the classical setup described in the previous chapter. For $1/2 < \kappa < 1$ it seems reasonable to expect that one of these, the Iwaniec–Rosser sieve [Iw3], might be optimal although this has not been proved. On the other hand, for $\kappa > 1$ the known results should not be expected to be best possible and quite conceivably are not even close.

Henceforth we shall therefore restrict ourselves to sequences $\mathcal{A}$ for which $\kappa = 1$, the "linear" sieve problems. This is by far the most important case and constituted three of the four examples given in the first chapter (all but the third example, for which we had $\kappa = 2$). This seems rather a nice circumstance since, all too frequently in mathematics, the case we would most like to know about is the most mysterious one. Here, the case which is by far the most important is, happily, also the one we know most about. However, the result of Bombieri shows that, even in the most favourable circumstances one cannot get primes, although one can come tantalizingly close. We should like to bridge this gap.

For $\kappa = 1$ the best possible functions F^+, F^-, first found by Jurkat and Richert [JR], may be defined as the continuous solutions of the differential-delay equations

$$\bigl(sF^{\pm}(s)\bigr)' = F^{\mp}(s-1),$$

together with the initial conditions

$$F^+(s) = \frac{2e^{\gamma}}{s}, \qquad F^-(s) = 0,$$

which hold for the starting interval $0 < s \leqslant 2$.

The Remainder Term

By about 1970 the theory of what one could or could not do with the main term in the linear sieve was already more or less developed to the extent it is today (although much of the foundational work of Iwaniec took another ten years to see the light of day). Important progress was about to shift to the nontrivial estimation of the remainder term

$$R(D) = \sum_{d \leqslant D} \lambda_d r_d.$$

Here we use λ_d to denote either λ_d^+ or λ_d^- and $r_d = r_d(x)$. Recall that what we called the "trivial" bound was the estimate

$$|R(D)| \leqslant \sum_{d \leqslant D} |\lambda_d r_d| \leqslant \sum_{d \leqslant D} |r_d|,$$

which in certain cases, such as example four, could be very nontrivial indeed. It is to such results that we refer when we speak of the "classical" sieve. Our main goal in this chapter is to see how one can improve on this bound.

We should remember that λ_d is similar to the Möbius function $\mu(d)$ and sometimes, as in our original Eratosthenian example, the remainder $R(D)$ turns out to be just as large as the main term. Certainly we can never take $D > x$ whether we are using the trivial bound or not. If we are using the trivial bound then we cannot even take $D > A(x)$. However, provided that we are not using the trivial bound then it is no longer obvious that we cannot take $D > A(x)$. Conceivably then we can go further in cases where $\mathcal{A}$ is "thin", that is $A(x)$ is quite small compared to x. But how can we accomplish this?

First let's return to our original example, that is the estimation of $\pi(x)$.

$$\mathcal{A} = \{m \leqslant x\}, \quad A_d(x) = \left[\frac{x}{d}\right] = \frac{x}{d} - \left\{\frac{x}{d}\right\}$$

$$A(x) = x, \quad g(d) = \frac{1}{d}, \quad r_d = -\left\{\frac{x}{d}\right\}.$$

In this very favourable situation we can get an admissibly small remainder even when we choose $D \approx x$. This is very good but on the other hand it is not good enough and moreover, impossible to improve on.

Now let's change the example a little and try instead to estimate the number of primes in the short interval $(x-y, x]$ where $y = x^\theta$ with $0 < \theta < 1$. Now $A(x) \approx y$, that is the situation is worse, so there is more room for improvement. Here, we have

$$A_d = \left[\frac{x}{d}\right] - \left[\frac{x-y}{d}\right] = \frac{y}{d} + r_d$$

where

$$r_d = \left\{\frac{x-y}{d}\right\} - \left\{\frac{x}{d}\right\} = \psi\left(\frac{x-y}{d}\right) - \psi\left(\frac{x}{d}\right).$$

Here, ψ is the "sawtooth" function $\psi(t) = t - [t] - 1/2$ which looks like

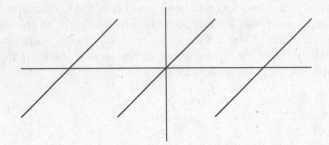

and has a very simple Fourier expansion:

$$\psi(t) = -\frac{1}{2\pi i} \sum_{h \neq 0} \frac{1}{h} e(ht), \qquad e(u) = e^{2\pi i u}.$$

Thus, our remainder term is given by $R(D) = R_x(D) - R_{x-y}(D)$ where

$$R_t(D) = \frac{1}{2\pi i} \sum_{\substack{h \in \mathbb{Z} \\ h \neq 0}} \frac{1}{h} S_h$$

and in turn,

$$S_h = \sum_{d \leqslant D} \lambda_d \, e\!\left(\frac{ht}{d}\right).$$

Now, $|r_d| \leqslant 1$ since it is the difference between the fractional parts of two numbers and hence the bound $|R(D)| \leqslant D$ follows trivially. This is also the trivial bound for the individual sum S_h. To do better it suffices to show, for both values of t and for each non-zero integer h, that S_h is small and, to get an improvement which will be useful, we need to beat by an essential amount (a fixed power of x), the above estimate $|R(D)| \leqslant D$.

The main term is approximately y (actually $y/\log x$) so we can take D almost as large as y. But, because $t \approx x$, the exponential factor $e(ht/d)$ varies in argument as d changes, even for larger d, namely those in the range $y < d < x^{1-\varepsilon}$. This range was empty for the original example where y was as large as x and this gives us hope to do better than the trivial bound.

There is a problem however in showing that the sum S_h is small. The exponential factor is not the only thing bouncing around. The coefficients λ_d, which after all are approximations to the Möbius coefficients $\mu(d)$, are also changing sign and in a not easily predictable fashion. How do we verify the (highly likely) proposition that these two effects are able to avoid nullifying each other?

Suppose we could somehow write $\{\lambda_d, \ d \leqslant D\}$ as a Dirichlet convolution $\lambda = \alpha * \beta$ where $\alpha = \{\alpha_m, \ m \leqslant M\}$, $\beta = \{\beta_n, \ n \leqslant N\}$, with $|\alpha_m| \leqslant 1$, $|\beta_n| \leqslant 1$ and $MN = D$. Thus $\lambda_d = \sum_{mn=d} \alpha_m \beta_n$ and

$$\sum_{d \leqslant D} \lambda_d r_d = \sum_{m \leqslant M} \sum_{n \leqslant N} \alpha_m \beta_n r_{mn}.$$

In the case of our example the sum S_h now becomes

$$S_h = \sum_{m \leqslant M} \sum_{n \leqslant N} \alpha_m \beta_n e\left(\frac{ht}{mn}\right),$$

and we are required to improve on the bound MN.

Of course the coefficients α, β are at least as mysterious as were the coefficients λ. As far as we are concerned they may as well be treated as if they were completely unknown bounded complex numbers. However, because we now have a double sum we can use Cauchy's inequality to rid ourselves of one of these two sets of unknown coefficients. For example, to dispense with the coefficients in the sum over m we may write

$$|S_h|^2 \leqslant \left(\sum_{m \leqslant M} |\alpha_m|^2 \right) \left(\sum_{m \leqslant M} \left| \sum_{n \leqslant N} \beta_n e\left(\frac{ht}{mn}\right) \right|^2 \right)$$

and now for this we are required to beat the estimate $M^2 N^2$. We can't hope to improve on the trivial bound M in the first sum and so we need to beat the bound MN^2 in the second one.

After an interchange of the order of summation the second sum becomes

$$\sum_{n_1 \leqslant N} \sum_{n_2 \leqslant N} \beta_{n_1} \overline{\beta}_{n_2} \sum_{m \leqslant M} e\left(\frac{ht}{m} \left(\frac{1}{n_1} - \frac{1}{n_2} \right) \right).$$

Here, in the inner sum there are no unknown coefficients! For the N pairs with $n_1 = n_2$ we cannot treat the inner sum non-trivially; the inner sum is M. However there are not so many of these pairs and their contribution MN to the double sum does beat MN^2. For the more generic pair $n_1 \neq n_2$ we have

$$\left| ht\left(\frac{1}{n_1} - \frac{1}{n_2} \right) \right| > \frac{x}{N^2} > M,$$

and so the exponential oscillates as m changes, provided that $MN^2 < x^{1-\delta}$.

In this case, using old ideas and results of van der Corput, the inner sum over m can be shown to have some cancellation and we do get an improvement. The conditions $x^{\theta+\delta} < MN$, $MN^2 < x^{1-\delta}$ are easily seen to be compatible for every $\theta < 1$ provided that we choose δ, M, and N wisely.

In fact, in modified form, the above arguments hold much more generally, and lead to many other applications.

But, how do we write λ as a convolution? There are now known to be a number of ways.

(A) The λ^2 decomposition

A decomposition of the required type was first accomplished by Motohashi [Mo]. He worked with the Selberg weights which (almost) decompose naturally as a product. More specifically we have, for every $m|P(z)$,

$$\sum_{d|m} \lambda_d^+ = \left(\sum_{d|m} \rho_d \right)^2$$

and so

$$\sum_d \lambda_d^+ r_d = \sum_{d_1|P(z)} \sum_{d_2|P(z)} \rho_{d_1} \rho_{d_2} r_{[d_1, d_2]}.$$

Now, the least common multiple is not quite a product but is almost so whenever the greatest common divisor is small and, since that is the case for most pairs d_1, d_2, this does not pose a serious problem.

A more substantial disadvantage of this approach is that we require $M = N$ in order to get a square and this lack of flexibility can limit the quality of the improvements.

(B) The Buchstab averaging

A second method of approach to this problem was subsequently employed by Chen [Ch] and since then by Friedlander–Iwaniec [FI1], Harman [Ha], Duke–Friedlander–Iwaniec [DFI], and others, and is based on the use of the Buchstab identity to replace a single remainder term by an average of remainder terms. Since we have

$$S(\mathcal{A}, z_2) = S(\mathcal{A}, z_1) - \sum_{z_1 \leqslant p < z_2} S(\mathcal{A}_p, p),$$

it follows that the employment of any sieve weights at all to each of the terms in the right hand sum leads to the remainder term

$$\sum_p \sum_d \lambda_d \, r_{dp}$$

which is exactly of the required form with $\alpha = \lambda$ and β being the characteristic function of the primes.

(C) The well-factorable weights

In 1977 Iwaniec [Iw4] gave a new choice of sieve weights which was a perturbation of the Iwaniec–Rosser weights. With these weights he was able to decompose the remainder term as a sum of (many, but not too many) terms, each of which factored into a bilinear form of the above type. Nevertheless the new weights were sufficiently close in shape to the original ones so as to leave the main terms in the upper and lower bounds essentially unchanged. This was vitally important, especially in the most important linear case $\kappa = 1$ where the Iwaniec–Rosser weights are the best possible.

An important feature of the Iwaniec weights is the flexibility that had been lacking in the Motohashi construction. Here there is no need to choose $M = N$ but rather we can take any $1 < M < D$, $MN = D$. Out of this resulted many further applications, one of the earliest being the result of Iwaniec that for infinitely many integers m the polynomial $m^2 + 1$ has at most two prime factors (the same result holding for the generic irreducible quadratic).

The proof of the sieve bounds for the above choice of weights is rather complicated. As is customary for sieves of combinatorial type, we always choose either $\lambda_d = \mu(d)$ or $\lambda_d = 0$ and the question becomes: When do we choose the one and when the other? In the original Iwaniec–Rosser weights, for an integer $d = p_1 \cdots p_r$ this choice depends on a set of inequalities of the type

$$p_1 \cdots p_j^{\beta+1} < D \text{ (or } > D) \text{ where } j \leqslant r, \ p_1 > p_2 > \cdots > p_r.$$

In the new "bilinear" weights we begin with a decomposition of r-dimensional space into boxes with edges $(P_j, P_j^{1+\delta})$. Corresponding to the above inequalities we consider instead those boxes which satisfy the inequalities

$$P_1 \cdots P_j^{\beta+1} < D \text{ (or } > D) \text{ where } j \leqslant r, \ P_1 > P_2 > \cdots > P_r.$$

and then attach to each such box the same choice ($\mu(d)$ or 0 as appropriate) for all integers $d = p_1 \cdots p_r$ with the r-tuple in the given box.

This modification has the effect of uncoupling the variables p_j without appreciably changing the sieve weights.

Combinatorial Identities

Following very little development up until the twentieth century the sieve of Eratosthenes has during the past hundred years grown substantially and in two somewhat distinct directions, beginning in the one case with the work of Brun and in the other with the work of Vinogradov. Some (but not all) of the leading figures in these two streams are given in the following chart.

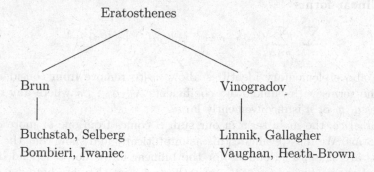

Although it is the direction initiated by Brun to which the words "sieve methods" are usually applied and which constitute the main theme of these notes, a very brief discussion of the latter area is certainly in order. Indeed, as the subject has developed in more recent years one sees that these two streams are re-approaching one another.

The methods of Vinogradov and his successors for the estimation of sums over primes begin with a decomposition of the von Mangoldt function $\Lambda(n)$ (or sometimes instead the Möbius function $\mu(n)$) judiciously as a sum of a small number of other functions, perhaps, just for illustration, thirteen of them: $f_1(n) + \cdots + f_{13}(n)$. We intend to study the same sum over primes

$$S = \sum_{n \leqslant x} a_n \Lambda(n)$$

as in Bombieri's sieve. This sum now inherits a decomposition

$$S = \sum_{j=1}^{13} S_j, \qquad S_j = \sum_{n \leqslant x} a_n f_j(n).$$

The hope is that, by a clever choice of this decomposition, the resulting sums are more easily dealt with, perhaps by using different techniques on the different constituents.

Using a combination of the elementary identities

$$\sum_{d|n} \Lambda(d) = \log n, \qquad \sum_{d|n} \mu(d) = 0,$$

one rearranges each of the sums S_j into one (or possibly both) of the following shapes:

(I) **linear forms**

$$\sum_{d \leqslant D} \lambda_d \, a_{md}, \qquad |\lambda_d| \leqslant 1,$$

and

(II) **bilinear forms**

$$\sum_{m \leqslant M} \sum_{n \leqslant N} \alpha_m \beta_n a_{mn}, \qquad |\alpha_m| \leqslant 1, \ |\beta_n| \leqslant 1.$$

The above elementary identities allow us to remove from consideration all of the terms with "unknown" coefficients λ_d, α_m, β_n where any of the variables d, m or n is inconveniently large.

In practice, the main term in our sum S comes from one or more of the linear forms. We must evaluate this asymptotically and show that the other subsums are small. The bounds for the bilinear forms don't depend on the nature of the coefficients α_m, β_n, just the fact that they are bounded. Note the similarity of all of this to the (more recent) sieve results described earlier in this chapter.

As compared to those sieve results the current method is more of a gamble. When it works it gives (usually) the asymptotics. Hence, one expects it to be less likely to work.

Sample combinatorial identity

We illustrate with just one of the many identities of this type. This particular one was discovered and applied originally by Linnik, see for example [Li], and has since been used successfully by Bombieri–Friedlander–Iwaniec [BFI], Heath-Brown [Hb1] (who had discovered it independently) and others.

For each integer $n > 1$,

$$\frac{\Lambda(n)}{\log n} = \sum_{j=1}^{\infty} \frac{(-1)^{j-1}}{j} t_j(n)$$

where $t_j(n)$ denotes the number of ways of writing n as the product of j integers each being strictly greater than one and with the order of the factors being distinguished (that is different orders all being counted). This reminds one of the ordinary divisor functions $\tau_j(n)$, the only difference being that in our case we insist that none of the factors be equal to one. Thus it is easy to see that τ_k can be expressed in terms of the functions t_j and so, by an elementary inversion formula it follows that

$$t_j(n) = \sum_{k=0}^{j} (-1)^{j-k} \binom{j}{k} \tau_k(n).$$

Thus, Linnik's identity reduces the study of $\Lambda(n)$ in a given sequence to the study in the same sequence of the various divisor functions $\tau_k(n)$. In the case where k is large and there are many factors it is possible in practice to arrange bilinear forms of the above type with great flexibility in the choice of M and N. It is then the smallest few values of k which present the limits to the quality of the results.

The proof of Linnik's identity is quite simple. We let $\zeta = \zeta(s)$ denote the Riemann zeta-function. We then have for each $j \geqslant 1$,

$$(\zeta - 1)^j = \sum_{n \geqslant 1} t_j(n) n^{-s},$$

so we deduce the result by comparing coefficients in the identity

$$\log \zeta = \log \left(1 + (\zeta - 1)\right) = \sum_{j=1}^{\infty} \frac{(-1)^{j-1}}{j} (\zeta - 1)^j,$$

where on the left side we use the Euler product formula for ζ and on the right side we use the Dirichlet series.

3 Primes of the form $X^2 + Y^4$

To begin with we ask the question "Which primes are the sum of two squares?" This is a very old and basic question which turns out to have a satisfyingly simple answer. In the case that the prime p is congruent to three modulo four it is easy to see that it is not the sum of two squares. Indeed, since squares of odd numbers are one modulo four and squares of even numbers are zero modulo four, no *integer* congruent to three modulo four can be the sum of two squares. Of course the prime two is the sum of two squares and that leaves us with the primes congruent to one modulo four. After trying a few examples one is led to the conjecture that every such prime may be so represented, but to prove it is a different story (although I did some years ago hear one mathematician, speaking with three hundred years of hindsight, declare it to be a triviality). Fermat stated that he had such a proof but it seems that the first recorded proof is due to Euler.

As a result of the combination of nineteenth century ideas of Dirichlet on primes in arithmetic progressions with those of Riemann, Hadamard and de la Vallée-Poussin, which gave the prime number theorem, we can even give an asymptotic formula:

$$\sum_{\substack{p \leqslant x \\ p = m^2 + n^2}} 1 \sim \frac{1}{2} \frac{x}{\log x}.$$

On the other hand, when it comes to polynomials of higher degree, at least those in a single variable, proofs are still lacking and we are only able to make conjectures.

Conjecture *For a certain positive constant c we have*

$$\sum_{\substack{p \leqslant x \\ p = m^2 + 1}} 1 \sim c \frac{x^{1/2}}{\log x}.$$

The set of integers of the form $m^2 + 1$ is a very thin set compared to the set of integers which are the sum of two squares and this has the effect of making their study very much more difficult. Until recently the thinnest polynomial sets which could be proved to represent infinitely many primes were those in a fairly generic family of quadratic polynomials in two variables, for example $m^2 + n^2 + 1$. Due to a result of Iwaniec [Iw2], we have for such a polynomial $\sum_{p \leqslant x} 1 \sim cx/(\log x)^{3/2}$.

More recently, during the year 1996 (published in 1998), Friedlander and Iwaniec [FI2, FI3, FI4] were able to successfully deal with a very much thinner set, the integers of the form $m^2 + n^4$. There we proved the expected asymptotic formula for the number of prime values up to x; this has the shape $\sum_{p \leqslant x} 1 \sim cx^{3/4}/\log x$.

We expect that the arguments extend (albeit, not without a good deal of hard work) to cover the case of primes of the form $\varphi(m, n^2)$ where φ is a general binary quadratic form. We did not however attempt to carry this out.

Still more recently, Heath-Brown [Hb2], using some similar ideas and also ideas of his own, was able to prove the expected asymptotic for primes of a still thinner set, those of the form $m^3 + 2n^3$, and Heath-Brown and Moroz [HM] have subsequently generalised that result to binary cubic forms for which, in the generic case, $\sum_{p \leqslant x} 1 \sim cx^{2/3} / \log x$.

We state more precisely the theorem of Friedlander and Iwaniec [FI3].

Theorem 1 *As m, n run through positive integers we have*

$$\sum\sum_{m^2+n^4 \leqslant x} \Lambda(m^2 + n^4) = \frac{4}{\pi} \kappa \, x^{3/4} \left\{ 1 + O\left(\frac{\log \log x}{\log x} \right) \right\},$$

where the constant is given by the elliptic integral

$$\kappa = \int_0^1 \left(1 - t^4 \right)^{1/2} \mathrm{dt} = \frac{\Gamma(1/4)^2}{6\sqrt{2\pi}}.$$

In this chapter we give an overview of the proof, showing how it fits into sieve framework as described earlier. As we recall, we study the sum

$$S(x) - \sum_{n \leqslant x} a_n \Lambda(n)$$

where $\Lambda(n)$, the von Mangoldt function, satisfies

$$\Lambda(n) = \sum_{d|n} \lambda_d, \qquad \lambda_d = -\mu(d) \log d,$$

so that

$$S(x) = \sum_{d \leqslant x} \lambda_d A_d(x),$$

where, as before,

$$A_d(x) = \sum_{\substack{n \leqslant x \\ n \equiv 0 \, (d)}} a_n.$$

For any $d \geqslant 1$ we postulate the basic approximation formula

$$(*) \qquad\qquad A_d(x) = g(d) A(x) + r_d(x),$$

and we need to make various assumptions about these quantities.

Assumptions

We divide these into three classes in accordance with the objects appearing in the previous formula.

(I) Assumptions about the counting functions

We begin with some crude bounds for $\mathcal{A} = (a_n)$, $a_n \geqslant 0$.

(I.1) $A(x) \gg A(\sqrt{x})(\log x)^2$.

(I.2) $A(x) \gg x^{1/3} \left(\displaystyle\sum_{n \leqslant x} a_n^2 \right)^{1/2}$.

(I.3) $A_d(x) \ll \dfrac{\tau(d)^8}{d} A(x)$, uniformly in $d \leqslant x^{1/3}$.

Note that in our case we have

$$a_n = \sum_{a^2+b^2=n} \mathfrak{Z}(b)$$

where $a, b \in \mathbb{Z}$, and $\mathfrak{Z}$ is the characteristic function on the set of squares, that is a_n is the number of representations of n as the sum of a square and a fourth power. In this example, as indeed rather generally, the above assumptions are not very difficult to check.

(II) Assumptions about the function g

We assume the following:

(II.1) g is a multiplicative function.

(II.2) $0 \leqslant g(p^2) \leqslant g(p) < 1$.

(II.3) $g(p^2) \ll p^{-2}$.

(II.4) $g(p) \ll p^{-1}$.

(II.5) $\displaystyle\sum_{p \leqslant y} g(p) = \log \log y + c + O\big((\log y)^{-10}\big)$ for some constant $c = c(g)$.

This last assumption means that we are dealing with the "linear" sieve. In our case the specific function g is given by

$$g(p) = \frac{1}{p} + \frac{\chi(p)}{p}\left(1 - \frac{1}{p}\right)$$

where $\chi(p) = \left(\dfrac{-1}{p}\right)$ is the Legendre symbol. As with the first set of assumptions, the verification of these axioms for our example, and for most other examples as well, does not provide any problems. The most difficult one (II.5) is essentially at the level of difficulty of the prime number theorem (more precisely the prime ideal theorem), with a relatively weak error term.

The constant 10 which occurs in the exponent in (II.5) is not important. It is large enough to suffice for the theorem and small enough that we can prove it for the application we have in mind. But then, the same could be said about 100. A similar remark applies to the constant 8 in (I.3).

(III) Assumption about the remainder term

First we state this in the precise form in which it is actually used in the work. We assume that the bound

$$\text{(III.1)} \qquad \sum_{d \leqslant DL^2}^{3} |r_d(t)| \leqslant A(x)L^{-2}$$

holds uniformly in $t \leqslant x$ for some $D > x^{2/3}$, where $\sum^3$ means that the sum runs over cubefree integers and where $L = (\log x)^{2^{24}}$.

That is certainly a rather technical looking assumption. Roughly speaking, we think of this as being the following:

We assume that the bound

$$\text{(R)} \qquad \sum_{d \leqslant D} |r_d(t)| \ll A(x)(\log x)^{-B}$$

holds uniformly in $t \leqslant x$, for some large B and some $D > x^{2/3}$.

Such an assumption is of just the type given in the first chapter, the standard remainder term assumption of the classical sieve. To prove (III.1) in the case of our example is not an easy task. The proof was given by Fouvry and Iwaniec [FvI] in connection with a different application. They achieved the bound with the choice $D = x^{\frac{3}{4}-\varepsilon}$, which is best possible apart from the ε. The fact that they got such a good level of distribution was a large part of the motivation for our beginning this project. As a result of their bound they deduced the expected asymptotic formula for the number of primes up to x which can be written as the sum of two squares one of which is the square of a prime.

(IV) Assumption on special bilinear forms

All of the above assumptions, or some variants thereof, are present in most works on the sieve. Our final assumption is rather different and is a new one. It is somewhat reminiscent of the bilinear form bounds discussed briefly in the second chapter in connection with the method begun by Vinogradov. A crucial distinction however is that we only require the successful treatment of forms having very special coefficients.

As in the case of the previous axiom (R) we first state this assumption in the precise form in which it will be needed. We assume

$$\text{(IV.1)} \qquad \sum_{m} \left| \sum_{\substack{N < n \leqslant 2N \\ mn \leqslant x \\ (n, m\Pi)=1}} \mu(n)\beta(n)a_{mn} \right| \leqslant A(x)(\log x)^{-2^{26}}$$

where

$$\beta(n) = \beta(n, C) = \sum_{\substack{c \mid n \\ c \leqslant C}} \mu(c).$$

This is required to hold for every C, $1 \leqslant C \leqslant x/D$, and every N having

$$\Delta^{-1}\sqrt{D} < N < \delta^{-1}\sqrt{x},$$

with some $\Delta \geqslant \delta \geqslant 2$ and some P in the range $2 \leqslant P \leqslant \Delta^{1/(2^{35} \log \log x)}$. Here $\Pi = \prod_{p<P} p$. Note that for $C = 1$ we have $\beta(n) = 1$.

As in the case of our axiom about the remainder term this is a rather complicated looking assumption so it is useful to give a rough (but therefore not completely accurate) way to look at it. Such a statement is achieved by making the following simplifications:

Think of $C = 1$ so that $\beta(n) = 1$.

Think of $P = 2$ so that $\Pi = 1$.

Think of $\delta = (\log x)^A$, $\Delta = x^\varepsilon$, where $A > 0$ is large and $\varepsilon > 0$ is small.

Then we require, for some large A, the inequality

(B) $$\sum_m \left| \sum_{\substack{N < n \leqslant 2N \\ mn \leqslant x}} \mu(n) a_{mn} \right| \leqslant A(x)(\log x)^{-A}$$

to hold, for every N with

$$x^{-\varepsilon}\sqrt{D} < N < (\log x)^{-A}\sqrt{x}.$$

Note that the more we know about the level D in assumption (R), that is the larger we can choose it, the less we need here in assumption (B). Thus, if in a given example we could take $D = x^{1-\varepsilon}$ then we would require this new axiom to hold only in a very narrow range with N near $\sqrt{x}$. In our case, since we have $D = x^{\frac{3}{4}-\varepsilon}$, we need to handle N all the way down to (a little below) $x^{3/8}$.

It is worth repeating that this is the new assumption that allows us to count the primes. To see why this additional property for our sequence might give us the chance to succeed let us recall the Selberg counter-example:

$$a_{mn} = \begin{cases} 1 & \nu(mn) \text{ even} \\ 0 & \nu(mn) \text{ odd} \end{cases}$$

where $\nu(r)$ counts the number of distinct prime factors of r. Here we see that, for each m, $\mu(n) a_{mn}$ has constant sign; thus (B) cannot hold. Hence this sequence no longer provides a counter-example and the detection of primes under this additional assumption cannot be ruled out.

Of course, being unable to prove something impossible is not the same as being able to prove that it *is* possible. So it could be the case that, even after this new axiom is included, it is still not possible to produce primes. However, this extra ingredient does turn out to make the difference. We have the following result.

Theorem 2 (Asymptotic sieve for primes) *Assume that the sequence* $\mathcal{A} = (a_n)$, $a_n \geqslant 0$ *satisfies assumptions* (I.1)–(I.3), (II.1)–(II.5), (III.1) *and* (IV.1). *Then*

$$\sum_{n \leqslant x} a_n \Lambda(n) = H A(x) \left\{ 1 + O \left(\frac{\log \delta}{\log \Delta} \right) \right\}$$

where

$$H = \prod_p \left(1 - g(p) \right) \left(1 - \frac{1}{p} \right)^{-1}$$

and the implied constant depends only on g.

We shall postpone a sketch of the proof of Theorem 2, given in [FI4], to the next chapter, concentrating for the remainder of this chapter on the derivation therefrom, given in [FI3], of Theorem 1.

Although Theorem 2 is quite a positive statement, it could conceivably be the case that it is of no practical value. We still need to show that there are interesting sequences for which the new axiom holds.

In fact, given Theorem 2 and the earlier remarks, the only thing remaining for the proof of Theorem 1 is to show that the new axiom (B), of course in its more precise form, holds for our sequence

$$a_n = \sum_{a^2 + b^2 = n} 3(b).$$

This however turns out to be a long and difficult task. In these notes we can hope only to draw attention to some of the highlights and try to give a rough feeling for what is going on.

Because of the assumptions we have made about the function g and due to the presence of the Möbius function it is possible to re-phrase our assumption (B) about the sums $\sum_m \left| \sum_n \mu(n) a_{mn} \right|$ so as to postulate instead a bound for the double sums $\sum_m \left| \sum_n \mu(n) r_{mn} \right|$. This latter sum is just a special case of what we had considered in the previous chapter, namely the general bilinear forms

$$\sum_m \alpha_m \sum_n \beta_n r_{mn}$$

where α_m, β_n are arbitrary bounded complex numbers.

It is of course the case that, if we could prove a corresponding statement about these general sums which is of the same strength as that in axiom (B), then the existence of primes in the given sequence would, a fortiori, follow. Indeed, the proof of Theorem 2 under this stronger assumption is much easier. However, it might well be more difficult (or even impossible) to prove this more general assumption holds for the sequence under consideration. Certainly, in the case of our example, we very much need in our proof to use the specific coefficients $\beta(n) = \beta(n, C)$ in our verification of the axiom (B).

Deduction of (B)

We want to bound the bilinear forms

$$\sum_m \sum_n \alpha_m \beta_n a_{mn}$$

in the special case where α_m denotes the absolute value, that is $\alpha_m = \text{sgn}\left(\sum_n \beta_n a_{mn}\right)$, and we think of β_n as being $\mu(n)$ even though in general it is only a close relative having similar properties, as described in axiom (IV.1).

As is very frequently the starting point in bounding such bilinear sums we want to apply Cauchy's inequality in the fashion described in the previous chapter. In our case however we need to specifically take α_m in the outer sum; more precisely we need to keep β_n inside so that we can later take advantage of the cancellation in the Möbius function μ. As is not hard to believe, for our given sequence a_n the arithmetic of the sequence is more natural in terms of the Gaussian integers $\mathbb{Z}[i]$ and it turns out to be important, if we are not to lose the game right at the start, to translate the problem into these terms before applying Cauchy.

We consider a_{mn}, the number of representations of mn as the sum of a square and a fourth power. If we write $w = u + iv$, $z = x + iy \in \mathbb{Z}[i]$, then we find that $\text{Re}\,\overline{w}z = ux + vy$, and

$$(u^2 + v^2)(x^2 + y^2) = (uy - vx)^2 + (ux + vy)^2.$$

Thus, our sequence becomes

$$a_{mn} = \sum_{|w|^2 = m} \sum_{|z|^2 = n} \mathfrak{Z}(\text{Re}\,\overline{w}z)$$

where, as we recall, $\mathfrak{Z}$ is the characteristic function on the set of squares.

Now, when we apply Cauchy's inequality we are led to consideration of the sum

$$\sum_{z_1} \sum_{z_2} \beta_{z_1} \overline{\beta}_{z_2} \mathcal{C}(z_1, z_2)$$

where

$$\mathcal{C}(z_1, z_2) = \sum_w f(w) \mathfrak{Z}(\text{Re}(\overline{w}z_1)) \mathfrak{Z}(\text{Re}(\overline{w}z_2))$$

or equivalently, as w runs over $\mathbb{Z}[i]$,

$$\mathcal{C} = \sum_{b_1} \sum_{b_2} f\big((b_1^2 z_2 - b_2^2 z_1)/\Delta\big).$$

Here f is a smooth weight function, which is introduced for technical reasons, and the double sum runs over pairs b_1, b_2 satisfying the congruence condition

$$b_1^2 z_2 \equiv b_2^2 z_1 \pmod{\Delta}$$

with $\Delta = \operatorname{Im} \overline{z_1} z_2$.

It is difficult to count these pairs since the modulus Δ is quite large; for given Δ the number of solutions is expected to be uniformly bounded. Nevertheless, we must try to do so. Fortunately we don't require the result to hold for each individual Δ. We begin by applying the Poisson summation formula in two variables, obtaining

$$\sum\sum_{h_1, h_2 \in \mathbb{Z}} G(h_1, h_2) F(h_1, h_2; z_1, z_2)$$

where F is a smooth function, and we then attempt to estimate the sum

$$G(h_1, h_2) = \sum_{\substack{\alpha_1 \ \alpha_2 \\ \alpha_1^2 z_2 \equiv \alpha_2^2 z_1 \pmod{\Delta}}} e\big((\alpha_1 h_1 + \alpha_2 h_2)\Delta^{-1}\big).$$

For each fixed pair $(h_1, h_2) \neq (0, 0)$ it is possible to show some cancellation on average over Δ, in other words over z_1, z_2, and this is all we require of these terms.

There still remains the 'main term' $h_1 = h_2 = 0$. Although we refer to it as the main term, since it comes from the zero frequency in the harmonic analysis, in fact this term too needs to be small. But now it is for reasons far more delicate.

This main term looks like

$$\sum_{z_1} \sum_{z_2} \beta_{z_1} \overline{\beta}_{z_2} \mathcal{C}_0(z_1, z_2)$$

where, for given z_1, z_2, the sum $\mathcal{C}_0(z_1, z_2)$ denotes the number of solutions of the congruence

$$\alpha_1^2 z_2 \equiv \alpha_2^2 z_1 \pmod{\Delta}.$$

Here we recall that $\alpha_1, \alpha_2 \in \mathbb{Z}$ and we are counting the solutions weighted by the smooth function f.

The latter count is given 'essentially' by the expression $\rho(z_2/z_1; \Delta)$ where $\rho(z; \Delta)$ counts the number of solutions of the congruence $\omega^2 \equiv z \pmod{\Delta}$, but only those in *rational* residue classes ω. More precisely, due to a co-primality problem, we also need to count the solutions modulo d for each divisor d of Δ. As is familiar, we can thus express ρ as a sum over the divisors of Δ of certain Jacobi symbols. Precisely,

$$G(0, 0) = \nu \sum_{\substack{d \mid \Delta \\ d \, \mathrm{odd}}} \frac{\varphi(d)}{d} \left(\frac{z_2/z_1}{d} \right)$$

where ν is defined by $2^\nu \| \Delta$.

If we insert this expression into $C_0(z_1, z_2)$ and then interchange the order of summation we are now led to sums of the type

$$S_\mathrm{I} = \sum_d \frac{\varphi(d)}{d} \sum_{z_1} \sum_{z_2} \atop {\Delta(z_1, z_2) \equiv 0\,(d)} f\left(\frac{|\Delta|}{d}\right) \beta_{z_1} \overline{\beta}_{z_2} \left(\frac{z_2/z_1}{d}\right).$$

This is split into sums $S_\mathrm{I}(D)$ in which the outer sum is over a dyadic range $D < d \leqslant 2D$ and then each subsum is given one of three treatments, depending on the size of D.

For the larger values of D, those with $D \geqslant \sqrt{x}$, we are going to begin by making the transformation $d \to |\Delta|/d$. This idea was first used by Dirichlet in his study of the divisor sum $\sum_{n \leqslant x} \tau(n)$. In our case the sum is much more complicated. It turns out that the law of quadratic reciprocity is required in an essential way in making this transformation and this leads to the replacement of the above sum S_I by the doubly twisted sum

$$S_\mathrm{II} = \sum_d \frac{\varphi(d)}{d} \sum_{z_1} \sum_{z_2} \atop {\Delta(z_1, z_2) \equiv 0\,(d)} f\left(\frac{|\Delta|}{d}\right) \beta_{z_1} \overline{\beta}_{z_2} \left(\frac{s_1}{r_1}\right)\left(\frac{s_2}{r_2}\right)\left(\frac{z_2/z_1}{d}\right).$$

Here the extra Jacobi symbols have the following meaning. We may write $z_j = r_j + i s_j$, where r is odd and s is even (since we are sieving for primes and so can assume the Gaussian integers of even norm have already been removed). It turns out that for z_1, z_2 as above, that is z_2/z_1 rational and such that $\Delta(z_1, z_2) \equiv 0\,(d)$, we have

$$\left(\frac{z_2/z_1}{d}\right) = \left(\frac{r_2/r_1}{d}\right) \quad \left(= \left(\frac{r_1 r_2}{d}\right)\right).$$

Moreover, the condition $\Delta \equiv 0 \pmod{d}$ is now replaced by the congruence

$$\overline{r}_1 s_1 \equiv \overline{r}_2 s_2 \pmod{d}$$

where, as usual, $\overline{r}$ denotes the multiplicative inverse of r modulo d. Hence, the above sums are of the form (ignoring the smooth function f)

$$S(D) = \sum_{D < d \leqslant 2D} \sum_{a\,(\mathrm{mod}\,d)} \left| \sum_{\overline{r}s \equiv a\,(\mathrm{mod}\,d)} \alpha_{rs}\left(\frac{r}{d}\right) \right|^2,$$

where in the case of S_I we have

$$\alpha_{rs} = \beta_z,$$

while in the case of S_II

$$\alpha_{rs} = \left(\frac{s}{r}\right)\beta_z.$$

To recapitulate, we need to treat for every D, $1 \leqslant D \leqslant x$, the sum $S_{\mathrm{I}}(D)$, saving from the trivial bound an arbitrary power of $\log x$. However, instead of treating $S_{\mathrm{I}}(D)$ in the range $x^{1/2} \leqslant D \leqslant x$, we can, due to the Dirichlet involution, instead treat $S_{\mathrm{II}}(D)$ in the range $1 \leqslant D \leqslant x^{1/2}$.

Case 1. The middle range

For the range

$$(\log x)^A \leqslant D \leqslant x(\log x)^{-A}$$

a treatment can be given which applies to very general coefficients and is reminiscent of the bounds that have been given for the generalized Barban sums

$$\sum_{D < d \leqslant 2D} \sum_{a \,(\mathrm{mod}\, d)} \left| \sum_{rs \equiv a \,(\mathrm{mod}\, d)} \gamma_r \delta_s \right|^2$$

in the same range.

The key differences here are that, in the first place we have $\bar{r}$ in place of r and in the second place we do not assume that there is any factorization of the coefficients in the form $\alpha_{rs} = \gamma_r \delta_s$. Nevertheless, using the above involution and after considerable work, the result does follow. As in the Barban case, it holds with quite general, indeed more general, coefficients α_{rs}.

Case 2. The range of small values

In this case we are dealing with the sums S_{I} in the range $D \leqslant (\log x)^A$. Here, we can prove the result holds for individual d, not merely on average over d. Unlike the previous case the result does however depend on the nature of the coefficients β_z, in particular their resemblance to the Möbius function. Indeed, one can write the Siegel–Walfisz theorem in the form

$$\sum_{\substack{m \leqslant x \\ m \equiv a \,(\mathrm{mod}\, q)}} \mu(m) \ll x(\log x)^{-A},$$

which is non-trivial only for $q \leqslant (\log x)^A$, and what we require (and obtain) is a generalization of this result to the setting of Hecke, rather than Dirichlet, characters.

Case 3. The range of large values

The final case, after the Dirichlet involution, reduces to the small values $D \leqslant (\log x)^A$ but now for the sum S_{II}. Here, due to the presence of the extra Jacobi symbol, the methods used for proving the generalized Siegel–Walfisz theorem do not apply.

This final case is the most difficult. We can no longer obtain the result for individual d as in case 2 and we can no longer prove the result for general coefficients as in case 1. The proof depends essentially on the nature of the coefficients. Not only are we required to obtain some cancellation from the presence of the Möbius function, but even more crucial is the cancellation coming from sums of the Jacobi symbols $\left(\dfrac{s}{r}\right)$ and $\left(\dfrac{r}{d}\right)$.

As an extra consequence of the cancellation in these latter sums we obtain the following result.

Given a prime $p \equiv 1 \pmod 4$ one can represent it in the form $p = r^2 + s^2$, and this representation is unique if we choose r and s to be positive with r odd. Let us define the 'spin' of the prime p to be the Jacobi symbol $\left(\dfrac{s}{r}\right)$. As a bonus result coming from our estimates in this part of the work we are able to show that, asymptotically, the spins of the primes are equally distributed between positive and negative. Precisely,

$$\sum_{\substack{p \leqslant x \\ p \equiv 1 \,(\mathrm{mod}\,4)}} \left(\frac{s}{r}\right) \ll x^{76/77}.$$

4 Asymptotic sieve for primes

In this final chapter we shall give a sketch of the proof of Theorem 2. Let $f : \mathbb{N} \to \mathbb{C}$ denote an arithmetic function and define the following truncations of f. Write

$$f_y^z(n) = \begin{cases} f(n) & \text{if } y < n \leqslant z, \\ 0 & \text{else.} \end{cases}$$

Denote $f^z = f_0^z$ in case $y = 0$, and $f_y = f_y^\infty$. Note that, for any z, we can decompose f as $f = f^z + f_z$.

We are going to decompose the von Mangoldt function. Beginning from the basic formula

$$\sum_{d|n} \mu(d) = \begin{cases} 1 & \text{if } n = 1 \\ 0 & \text{if } n > 1, \end{cases}$$

we deduce that

$$\Lambda_z(n) = \sum_{bc|n} \sum \mu(b)\Lambda_z(c)$$

$$= \sum_{bc|n} \sum \mu^y(b)\Lambda_z(c) + \sum_{bc|n} \sum \mu_y(b)\Lambda_z(c).$$

In the first sum replace Λ_z by $\Lambda - \Lambda^z$. Using the second basic formula

$$\sum_{d|n} \Lambda(d) = \log n,$$

we obtain

$$\Lambda_z(n) = \sum_{b|n} \mu^y(b) \log \frac{n}{b} - \sum_{bc|n}\sum \mu^y(b)\Lambda^z(c) - \sum_{bc|n}\sum \mu_y(b)\Lambda_z(c).$$

Here the key fact emerging from these manipulations is that in the first sum on the right the only variable b is small, in the second sum both variables b and c are small, while in the third they are both large. The identity produced here is quite similar to a well-known identity of Vaughan [Va].

We want to consider the sum

$$\sum_{n \leqslant x} a_n \Lambda(n) = \sum_{n \leqslant x} a_n \Lambda_z(n) + O(A(z)\log z).$$

This first error is quite acceptable. Think of z as being near $\sqrt{x}$; in practice it will be even smaller.

Inserting in this our just-completed decomposition of Λ_z we obtain the corresponding decomposition:

$$\sum_{n \leqslant x} a_n \Lambda_z(n) = S_1 - S_2 - S_3.$$

Estimation of S_1

We have

$$S_1 = \sum_{n \leqslant x} a_n \sum_{b|n} \mu^y(b) \log \frac{n}{b} = S_{11} - S_{12},$$

where

$$S_{11} = \sum_{b \leqslant y} \mu(b) \sum_{\substack{n \leqslant x \\ n \equiv 0 \,(\mathrm{mod}\, b)}} a_n \log n,$$

and

$$S_{12} = \sum_{b \leqslant y} \mu(b) \log b \sum_{\substack{n \leqslant x \\ n \equiv 0 \,(\mathrm{mod}\, b)}} a_n.$$

The sum S_{11}

By partial summation we have

$$\sum_{\substack{n \leqslant x \\ n \equiv 0 \,(\mathrm{mod}\, b)}} a_n \log n = A_b(x) \log x - \int_1^x A_b(t) \, \frac{dt}{t}.$$

We introduce once again our basic approximation formula

$$(*) \qquad\qquad A_b(t) = A(t)g(b) + r_b(t).$$

The main term of $(*)$ when inserted in S_{11} gives a contribution

$$M_{11} = A(x) \log x \sum_{b \leqslant y} \mu(b) g(b) - \left(\int_1^x A(t) \frac{dt}{t} \right) \sum_{b \leqslant y} \mu(b) g(b).$$

Now, we have

$$\sum_{b \leqslant y} \mu(b) g(b) \ll (\log y)^{-A},$$

say with $A = 9$. This follows, using familiar techniques, from assumption (II.5), which is essentially equivalent to the prime number theorem. Thus, as long as $y > x^\varepsilon$ the insertion of this gives an admissible bound for M_{11}.

On the other hand, the remainder term from $(*)$ gives to S_{11} a contribution R_{11} which satisfies the bound

$$R_{11} \ll (\log x) \sup_{t \leqslant x} \sum_{b \leqslant y} |r_b(t)|$$

$$\ll A(x) (\log x)^{1-A},$$

the latter inequality following from (R) provided that $y \leqslant D$.

The sum S_{12}

In this sum as well, we introduce the approximation $(*)$. The main term of $(*)$ when inserted into S_{12} gives the expected main term in the theorem:

$$M_{12} = A(x) \sum_{b \leqslant y} \mu(b) g(b) \log b$$

$$= -H A(x) + O\big(A(x) (\log x)^{-A} \big),$$

provided that $y > x^\varepsilon$, this last step following from (II.5) and the identity

$$- \sum_{b=1}^\infty \mu(b) g(b) \log b = H,$$

which may in turn be deduced under our assumptions about the function g.

The remainder term from $(*)$ gives to S_{12} a contribution R_{12} which satisfies

$$R_{12} \ll (\log x) \sum_{b \leqslant y} |r_b(x)| \ll A(x) (\log x)^{1-A}.$$

Again, just as with the previous remainder term R_{11}, this follows from (R) under the same condition, $y \leqslant D$.

Estimation of S_2

Recall that our second sum is given by

$$S_2 = \sum_{n \leqslant x} a_n \sum_{bc|n} \mu^y(b) \Lambda^z(c)$$

$$= \sum_{b \leqslant y} \mu(b) \sum_{c \leqslant z} \Lambda(c) \sum_{\substack{n \leqslant x \\ n \equiv 0 \,(\mathrm{mod}\, bc)}} a_n$$

and, by $(*)$, we can write this as

$$S_2 = \sum_{b \leqslant y} \mu(b) \sum_{c \leqslant z} \Lambda(c)\{A(x)g(bc) + r_{bc}(x)\}.$$

We are now going to make, for the purposes of this sketch of the proof, the technically simplifying assumption that in the above sum we can pretend that $g(bc) = g(b)g(c)$. This certainly happens if our function g is completely multiplicative or if our sequence is supported on squarefree numbers, but this is not quite the case in our example. That causes only a little difficulty in making things precise; however, under this simplifying assumption we see at once that the main part of the above sum is just

$$M_2 = A(x) \left(\sum_{b \leqslant y} \mu(b)g(b) \right) \left(\sum_{c \leqslant z} \Lambda(c)g(c) \right).$$

Here, the first sum satisfies

$$\sum_{b \leqslant y} \mu(b)g(b) \ll (\log y)^{-A}$$

and, as already noted in the estimation of M_{11}, the second one is bounded by

$$\sum_{c \leqslant z} \Lambda(c)g(c) \ll \log z,$$

using (II.4) and the Chebyshev bound. On the other hand, the remainder term in $(*)$ gives a contribution to S_2 which is

$$R_2 = \sum_{b \leqslant y} \mu(b) \sum_{c \leqslant z} \Lambda(c)\, r_{bc}(x)$$

so that

$$|R_2| \leqslant \log z \sum_{d \leqslant yz} \tau(d)\, |r_d(x)|.$$

This is not quite in the form required for an application of the axiom (R) because of the presence of the divisor function $\tau(d)$. In fact however, using (R) together with the other assumptions, it is possible to show that this modified sum satisfies the same bound weakened only by a few logarithms so that we have, say,

$$R_2 \ll A(x)(\log x)^{10-A},$$

provided that $yz \leqslant D$.

In practice we shall make the choice $y = z = \Delta^{-1}\sqrt{D}$ where, Δ is a parameter mentioned in axiom (IV.1) and in the statement of Theorem 2. The accompanying parameter δ will arise very soon.

Estimation of S_3

We next consider our final sum, by far the most complicated to deal with:

$$S_3 = \sum_{n \leqslant x} a_n \sum\sum_{bc|n} \mu_y(b)\Lambda_z(c).$$

Recalling our notation, this means that both variables b and c are rather large. Suppose, if possible, that y and z could be chosen so large that $yz > x$. Then, for any integer n contributing to S_3, we would have $yz \leqslant bc \leqslant n \leqslant x$. This is impossible to satisfy so we would then have an empty set and $S_3 = 0$.

Thus, if we could choose $yz > x$ we would be done. Unfortunately, life is not so easy. Recall that in order to use assumption (R) in our treatment of S_2 we required $yz \leqslant D$ and we can never take $D > x$ in (R). Hence, in any realistic application there will always be some terms in S_3.

Our next best hope is that in some very lucky cases our sequence might have the following two very favourable properties:

(I) $\mathcal{A}$ almost dense, that is $A(x) > x^{1-\varepsilon}$.

(II) $\mathcal{A}$ has very good distribution in progressions, that is $D > \left(A(x)\right)^{1-\varepsilon}$.

Under these circumstances there will be relatively few terms in S_3 and so we can hope to bound their contribution as an error term. This actually works in certain cases, even without resorting to any arguments that detect cancellation in the error terms. For example, for quite general sequences Bombieri's sieve gives, under these conditions, asymptotic results for the sum $\sum_{n \leqslant x} a_n \Lambda_k(n)$ for every integer $k \geqslant 2$. However, it just barely fails to do so in the most crucial case $k = 1$ which would have then shown that the sequence contains primes. It is at exactly this point where our new axiom (B) comes to the rescue.

Decomposition of S_3

We let $\sigma > 1$ be a parameter to be chosen and we decompose

$$S_3 = \sum_{b>y} \mu(b) \sum_{c>z} \Lambda(c) \sum_{\substack{n \leqslant x \\ n \equiv 0 \,(bc)}} a_n$$

$$= \sum_{b>\sigma y} \sum_{c>\sigma z} + \sum_{\sigma y \geqslant b > y} \sum_{c>z} + \sum_{b>\sigma y} \sum_{\sigma z \geqslant c > z}$$

$$= S_{31} + S_{32} + S_{33}, \quad \text{say.}$$

The sum S_{31}

Here we have $bc > \sigma^2 yz$. We shall choose σ so that $\sigma^2 yz > x^{1-\varepsilon(x)}$ for some function $\varepsilon(x) \to 0$ as $x \to \infty$. Then, just as before, there are not many b, c so that S_{31} can be bounded trivially as an error term. As before we require $yz \leqslant D$ (or else we would run into hopeless problems with S_2). But this no longer implies that $D > x^{1-\varepsilon}$, due to the presence of σ. We still could not require $\sigma^2 yz > x$ which would make S_{31} empty. In this case we would be able to prove (rather more easily) the sieve result but there would be no sequences to which it would apply.

In practice we choose

$$\sigma = \frac{\Delta}{\delta}\sqrt{\frac{x}{D}}$$

so that

$$\sigma y = \sigma z = \delta^{-1}\sqrt{x},$$

and we think of this latter quantity as being $\sqrt{x}/(\log x)^A$.

Here we expect to make a saving over the trivial estimate which is entirely engendered by the fact that the variables b and c are restricted to a narrow range and by the above choices it follows that the amount of saving is controlled by δ.

There is a technical point here which we briefly discuss but whose precise implementation we shall omit from this sketch. When we make a trivial estimate of the sum S_{31} as described above, although we make a saving related to δ we actually lose a factor $\log x$ which ruins everything. This forces us to reconsider the whole proof ab initio. Instead of beginning with our original sequence $\mathcal{A}$ we start with the sequence

$$\mathcal{A}^* = (a_n^*), \quad n \leqslant x,$$

where

$$a_n^* = \begin{cases} a_n & \text{if } (n, P(w)) = 1 \\ 0 & \text{else,} \end{cases}$$

with a small parameter w. Note that the contribution from primes to $\mathcal{A}^*$ is exactly the same as it was to $\mathcal{A}$ apart from the very small amount coming from primes not exceeding w. However, it turns out that, when we carry out the same computations for the sequence $\mathcal{A}^*$ as we have sketched for $\mathcal{A}$, everything works just about as before except that now, on reaching the point of making a trivial estimate for S_{31}, we no longer lose that logarithmic factor.

In carrying out these modified computations we use the fact that, as noted in the course of the first chapter (see the first diagram), it is possible to give sieves which are very accurate when we do not attempt to sieve to a very high level. As a result, taking w to be fairly small we can describe the new sequence $\mathcal{A}^*$ roughly speaking in terms of a sieve, that is, on average over n

$$a_n^* \approx a_n \sum_{d \mid n} \lambda_d,$$

for a certain choice of sieve weights $\{\lambda_d;\ d \leqslant \Delta,\ d\,|\,P(w)\}$. We say that we have subjected the sequence $\mathcal{A}$ to a "preliminary sieve" and it is when $\mathcal{A}^*$ is written in this form that we can perform the required computations. These are then extremely similar to those for $\mathcal{A}$ apart from the more cumbersome notation.

The result achieved is a bound

$$S_{31} \ll A(x) \frac{\log \delta}{\log \Delta}$$

which is much worse than the bounds for all the other error terms and is barely sufficient to give the result of Theorem 2.

The above technique has been around a long time in one form or another, possibly first used by Linnik, but it is difficult to give a definitive attribution. Certainly, in the current context it is especially appropriate to cite Bombieri's work [Bo2] on the Asymptotic Sieve and it is also subsequently to be found in [BFI] and [Hb1] as well as other places.

Enter the bilinear forms

It still remains to estimate the sums S_{32} and S_{33} and it is for these alone that we need the bilinear form axiom (B); actually in one case we require it in its more precise form (IV.1). Having confessed our sin of treating the earlier sums for the slightly simpler sequence $\mathcal{A}$ rather than for $\mathcal{A}^*$, which was really required in order to succeed with S_{31}, we shall now switch over to the treatment of $\mathcal{A}^*$ for the final two sums. There is only a tiny difference in dealing with one rather than the other.

The sum S_{32}

In the case of $\mathcal{A}^*$ this sum is of the form

$$S_{32} = \sum_m \lambda_m \sum_{c>z} \Lambda(c) \sum_{y<b\leqslant\sigma y} \mu(b)\, a_{bcm}.$$

We write $k = cm$, and use again the basic formula

$$\sum_{c|k} \Lambda(c) = \log k.$$

This gives the bound

$$|S_{32}| \leqslant \log x \sum_k \left| \sum_{y<b\leqslant\sigma y} \mu(b)\, a_{bk} \right|.$$

Here, after making a dyadic subdivision of the range of summation over b we see that the double sum is bounded by $\ll \log \sigma$ copies of the sum occurring in axiom (B). Since $\sigma \leqslant x$ we deduce that

$$S_{32} \ll A(x)(\log x)^{2-A}$$

as a consequence of axiom (B), the special case $C = 1$ of axiom (IV.1).

The sum S_{33}

Here we have

$$S_{33} = \sum_r \lambda_r \sum_{b > \sigma y} \mu(b) \sum_{z < c \leqslant \sigma z} \Lambda(c) \, a_{bcr}.$$

This is rather similar to S_{32}, but now it is the variable c rather than b which is well-located. This causes us a problem. Because here the inner sum $\sum_c \Lambda(c) a_{bcr}$ does not change sign we cannot simply insert absolute value signs and proceed as before. First there has to be some kind of rearrangement of the sum.

One solution to this problem is provided by the following. Since the von Mangoldt function Λ is built out of sums of the Möbius function μ we can hope to split up and rearrange the inner sum before inserting absolute values so as to obtain inner sums which have a reasonable expectation of cancellation.

To fix ideas we shall make the simplifying assumption that c is squarefree; in fact because of the fact that c occurs as $\Lambda(c)$ so is supported on prime powers the contribution from non-squarefree c is very small.

For all $c > 1$ which are squarefree we have the formula

$$\mu(c)\Lambda(c) = -\sum_{m \mid c} \mu(m) \log \frac{C}{m},$$

which is valid for every $C > 0$.

We take in particular $C = x D^{-1}$ as in axiom (IV.1). Thus

$$\Lambda(c) = \mu^2(c)\Lambda(c) = X_1(c) - X_2(c),$$

say, where

$$X_1(c) = \sum_{\substack{m \mid c \\ m < C}} \mu(m) \log \frac{C}{m},$$

$$X_2(c) = \sum_{\substack{m \mid c \\ m > C}} \mu(m) \log \frac{C}{m}.$$

Contribution to S_{33} from X_1

It is easy to check that we can express X_1 by the following integral:

$$X_1(c) = \int_1^C \beta(c, t) \frac{dt}{t},$$

where $\beta(c, t)$ is the coefficient from axiom (IV.1). Hence, the contribution from X_1 to S_{33} is given by

$$S_{331} = \sum_r \lambda_r \sum_{b>\sigma y} \mu(b) \sum_{z<c\leqslant\sigma z} \int_1^C \beta(c,t)\, \frac{dt}{t}\, a_{bcr}.$$

We interchange the order and write $m = br$ getting

$$S_{331} = \int_1^C \sum_m \tau(m) \left| \sum_c \beta(c,t)\, a_{mc} \right| \frac{dt}{t}.$$

Here the larger range for the variable c requires us to make, for each t, a number of applications of axiom (IV.1) ($\ll \log x$ of them in number). More precisely, here we require a modified form of the axiom weighted by the divisor function $\tau(m)$ as was earlier the case for axiom (R). As there the weighted version of (IV.1) can be deduced from the unweighted version in the presence of the other axioms. The final result of this again loses a few unimportant logarithms leading to the bound

$$S_{331} \ll A(x)(\log x)^{12-A}.$$

Contribution to S_{33} from X_2

All that remains is to prove that we can obtain a similar bound for the contribution to S_{33} coming from X_2. That contribution is

$$S_{332} = \sum_r \lambda_r \sum_{b>\sigma y} \mu(b) \sum_{z<c\leqslant\sigma z} a_{bcr} \sum_{\substack{m|c \\ m>C}} \mu(m) \log \frac{C}{m}.$$

We are going to remove the logarithm here; in fact we can do so by replacing it with an integral $\int dt/t$ and then interchanging the order.

Now, set $c = m\ell$. Note that $n = rbm\ell \leqslant x$, and $m > C = xD^{-1}$, so that $rb\ell \leqslant D$.

We can write our sum in the form

$$S_{332} = \int_C^x \left(\sum_r \lambda_r \sum_\ell \mu(\ell) \sum_b \mu(b) \sum_m a_{r\ell bm} \right) \frac{dt}{t}.$$

Here the inner sum over m, by the approximation $(*)$ looks like:

$$\sum_m = g(d) \sum_{n\in I} a_n + r_d(w_2) - r_d(w_1)$$

where $d = r\ell b$ and where $I = (w_1, w_2]$ is an interval whose endpoints depend on r, ℓ and b.

Recall that $d \leqslant D$. Thus, the contribution of the two remainders to the above integral S_{332} is small by axiom (R). Actually we require the modified (R) with divisor function weights because of the decomposition $d = r\ell b$.

For the main term in the above integral for S_{332} we fix r and ℓ and sum over b. This leads us once again to the sum $\sum_b \mu(b)g(b)$ but over a range where

b is fairly large. Due to the condition $b > \sigma y$, this sum is small by axiom (II.5), the one which generalizes the prime number theorem. As a result we can then sum trivially over the variables r and ℓ.

Combining the results of the previous two paragraphs we obtain a bound for S_{332}. Because of the specific exponent 10 in axiom (II.5) (which was chosen just for illustration and could easily be strengthened) we obtain a not very strong bound:

$$S_{332} \ll A(x)(\log x)^{-5}.$$

This is however more than sufficient and is in any case better than the bound for S_{31} which sets the limit to the error term in the statement of the theorem.

Finally, having successfully estimated all of the sums S_j, $1 \leqslant j \leqslant 3$, in our decompositon, we complete the proof of Theorem 2.

5 Conclusion

One cannot help but feel that there has been a lot of progress with sieve methods during the past century. It is particularly exciting that the original elementary ideas which gave birth to the subject can now be fruitfully combined with the tools of analytic number theory and also, via the exploitation of exponential sums, with arithmetic geometry. In this way one not only brings into play many of the deepest ideas from various branches of number theory but uses them in a seemingly essential way to achieve the final goal of the method, the successful detection of primes in arithmetically interesting sequences. Now that the parity problem is not the insurmountable obstacle it once seemed to be, it is easy to be more optimistic about the future progress in the subject. On the other hand, it is also easy to get a little discouraged when our favourite sequences are still awaiting attention and when one sees how very much work is required for each of the few examples which have been successfully treated so far.

Certainly though the future is bright, as time will show. How much time? Well, that is another question.

References

[Bo1]   E. Bombieri, *On the large sieve*, Mathematika 12 (1965), 201-225.

[Bo2]   E. Bombieri, *The asymptotic sieve*, Mem. Accad. Naz. dei XL 1/2 (1975/76), 243-269.

[BFI]   E. Bombieri, J. B. Friedlander and H. Iwaniec, *Primes in arithmetic progressions to large moduli II*, Math. Ann. 277 (1987), 361-393; *III*, J. Amer. Math. Soc. 2 (1989), 215-224.

[Br]   V. Brun, *Le crible d'Eratosthène et le théorème de Goldbach*, C. R. Acad. Sci. Paris 168 (1919), 544-546.

[Bu]   A. A. Buchstab, *New improvements in the method of the sieve of Eratosthenes*, Mat. Sb. 46 (1938), 375-387 (Russian).

[Tc]   P. L. Chebyshev, *Sur la fonction qui détermine la totalité des nombres premiers inferieurs à une limite donnée*, J. Math. Pures et Appl. 17 (1852), 366-390.

[Ch]   J.-R. Chen, *On the distribution of almost primes in an interval*, Sci. Sinica 18 (1975), 611-627.

[DFI]   W. Duke, J. B. Friedlander and H. Iwaniec, *Equidistribution of roots of a quadratic congruence to prime moduli*, Ann. of Math. 141 (1995), 423-441.

[Fd]   K. Ford, *On Bombieri's asymptotic sieve*, Trans. Amer. Math. Soc. 357 (2005), 1663-1674.

[FvI]   E. Fouvry and H. Iwaniec, *Gaussian primes*, Acta Arith. 79 (1997), 249-287.

[FI1]   J. B. Friedlander and H. Iwaniec, *Quadratic polynomials and quadratic forms*, Acta Math. 141 (1978), 1-15.

[FI2]   J. B. Friedlander and H. Iwaniec, *Using a parity-sensitive sieve to count prime values of a polynomial*, Proc. Nat. Acad. Sci. U.S.A. 94 (1997), 1054-1058.

[FI3]   J. B. Friedlander and H. Iwaniec, *The polynomial $X^2 + Y^4$ captures its primes*, Ann. of Math. 148 (1998), 945-1040.

[FI4]   J. B. Friedlander and H. Iwaniec, *Asymptotic sieve for primes*, Ann. of Math. 148 (1998), 1041-1065.

[Gr]   G. Greaves, *Sieves in number theory*, Ergeb. der Math. vol. 43, Springer-Verlag, Berlin, 2001.

[HR]   H. Halberstam and H.-E. Richert, *Sieve methods*, London Math. Soc. Monographs vol. 4, Academic Press, London, 1974.

[Ha]   G. Harman, *On the distribution of αp modulo one*, J. London Math. Soc. 27 (1983), 9-18.

[Hb1]   D. R. Heath-Brown, *The number of primes in a short interval*, J. Reine Angew. Math. 389 (1988), 22-63.

[Hb2]   D. R. Heath-Brown, *Primes represented by $x^3 + 2y^3$*, Acta Math. 186 (2001), 1-84.

[HM]   D. R. Heath-Brown and B. Z. Moroz, *Primes represented by binary cubic forms*, Proc. London Math. Soc. 84 (2002), 257-288.

[Iw1]   H. Iwaniec, *On the error term in the linear sieve*, Acta Arith. 19 (1971), 1-30.

[Iw2]   H. Iwaniec, *Primes represented by quadratic polynomials in two variables*. Bull. Acad. Polon. Sci. Ser. Sci. 20 (1972), 195-202.

[Iw3]   H. Iwaniec, *Rosser's sieve*, Acta Arith. 36 (1980), 171-202.

[Iw4]   H. Iwaniec, *A new form of the error term in the linear sieve*, Acta Arith. 37 (1980), 307-320.

[Iw5]   H. Iwaniec, *Sieve methods*, Rutgers University lecture notes, New Brunswick, 1996.

[JR]   W. B. Jurkat and H.-E. Richert, *An improvement of Selberg's sieve method I*, Acta Arith. 11 (1965), 217-240.

[Li]   Yu. V. Linnik, *The dispersion method in binary additive problems*, Transl. Math. Monographs vol. 4, Amer. Math. Soc., Providence, 1963.

[Mo]   Y. Motohashi, *On some improvements of the Brun-Titchmarsh theorem*, J. Math. Soc. Japan 26 (1974), 306-323.

[Se1]   A. Selberg, *On an elementary method in the theory of primes*, Norske Vid. Selsk. Forh. Trondheim 19 (1947), 64-67.

[Se2]   A. Selberg, *On elementary methods in primenumber-theory and their limitations*, Den 11te Skand. Matematikerkong., Trondheim (1949), 13-22; Johan Grundt Tanums Forlag, Oslo, 1952.

[Se3]   A. Selberg, *Sieve methods*; 1969 Number Theory Institute, SUNY, Stony Brook, 311-351; Proc. Sympos. Pure Math. vol. 20, Amer. Math. Soc., Providence, 1971.

[Va]   R. C. Vaughan, *Sommes trigonométriques sur les nombres premiers*, C. R. Acad. Sci. Paris Ser. A 285 (1977), 981-983.

[Vi]   I. M. Vinogradov, *Selected works*, Springer-Verlag, Berlin, 1985.

Counting Rational Points on Algebraic Varieties

D. R. Heath-Brown

Mathematical Institute, University of Oxford
24-29 St Giles, Oxford OX1 3LB, England
e-mail: rhb@maths.ox.ac.uk

1 First lecture. A survey of Diophantine equations

1.1 Introduction

In these lectures we will be interested in solutions to Diophantine equations $F(x_1, \ldots, x_n) = 0$, where F is an absolutely irreducible polynomial with integer coefficients, and the solutions are to satisfy $(x_1, \ldots, x_n) \in \mathbb{Z}^n$. Such an equation represents a hypersurface in $\mathbb{A}^n$, and we may prefer to talk of integer points on this hypersurface, rather than solutions to the corresponding Diophantine equation. In many cases of interest the polynomial F is homogeneous, in which case the equation defines a hypersurface in $\mathbb{P}^{n-1}$, and the non-zero integer solutions correspond to rational points on this hypersurface. In this situation the solutions of $F(x_1, \ldots, x_n) = 0$ form families of scalar multiples, and each family produces a single rational point on the corresponding projective hypersurface. Occasionally we shall encounter systems of two or more equations, and these may correspond to varieties of codimension 2 or more, rather than hypersurfaces.

Much work in the theory of Diophantine equations has been directed at showing that certain classes of equations have finitely many solutions. However we shall be interested in those cases where either we expect the number of solutions to be infinite, or we expect the number to be finite but cannot prove it. In these cases it is sensible to ask for bounds on the number of solutions which might lie in a large region $\max |x_i| \leqslant B$, say.

1.2 Examples

Let us look at some examples.

1. The equation

$$x_1^k + x_2^k = x_3^k + x_4^k. \tag{1.1}$$

It is expected that there are only the trivial solutions as soon as $k \geqslant 5$, but since we are unable to prove this, one may ask for an upper bound on the number of non-trivial solutions in a given large box.

2. The equation

$$x_1^k + x_2^k + x_3^k = N, \quad x_1, x_2, x_3 \geqslant 0.$$

Here it is believed that there is at most one solution, up to permutation, as soon as $k \geqslant 7$, but in the absence of a proof we ask for upper bounds on the number of solutions. (When $k \leqslant 6$ we know of infinitely many essentially different examples in which N has two or more representations.)

3. In Waring's problem one encounters the equation

$$x_1^k + \ldots + x_s^k = x_{s+1}^k + \ldots + x_{2s}^k, \quad 0 \leqslant x_1, \ldots, x_{2s} \leqslant B.$$

If one can show that there are $O(B^{2s-k+\varepsilon})$ solutions, for any fixed $\varepsilon > 0$, one can deduce the Hardy–Littlewood asymptotic formula for representations of a large integer N as a sum of $2s + 1$ perfect k-th powers. Thus one would have $G(k) \leqslant 2s + 1$, providing s is large enough for the usual local conditions to hold.

4. Vinogradov's Mean Value Theorem relates to the system of equations

$$x_1^h + \ldots + x_s^h = x_{s+1}^h + \ldots + x_{2s}^h, \quad (1 \leqslant h \leqslant k),$$

in which $0 \leqslant x_1, \ldots, x_{2s} \leqslant B$. It is known that if s is sufficiently large, then the number of solutions is $O(B^{2s-k(k+1)/2})$. (In fact this is known to hold if $s \geqslant \{1 + o(1)\}k^2 \log k$.) Such bounds have numerous applications, for example to estimates for the zero-free region of the Riemann Zeta-function. One could conjecture that the same bound holds as soon as $s > k(k + 1)/2$. If true, this would lead to improved results on the Zeta-function.

5. Manin's conjecture. As a simple special case of Manin's conjecture, let $F(x_1, x_2, x_3, x_4)$ be a non-singular[1] cubic form with integral coefficients, and suppose that there is at least one non-zero integral solution to the equation

$$F(x_1, x_2, x_3, x_4) = 0.$$

Then the conjecture states that the number of non-trivial solutions in the box $\max |x_i| \leqslant B$ will be asymptotically $cB(\log B)^r$ for a suitable positive constant c, where r is the rank of the Picard group of the surface $F = 0$. (This is not quite the usual formulation, since we have not insisted that our solutions should be projectively distinct.) No non-singular cubic surface

---

[1] A form in n variables will be said to be non-singular if $\nabla F(\mathbf{x})$ is non-zero for every non-zero $\mathbf{x} \in \overline{\mathbb{Q}}^n$.

is known for which Manin's conjecture can be established. Indeed, even the weaker statement that the number of non-trivial solutions is $O(B^{1+\varepsilon})$ for every $\varepsilon > 0$, eludes us.

6. For $D > 0$ the equation

$$x^2 + Dy^2 = z^3, \quad 1 \leqslant z \leqslant D^{1/2},$$

may be used to count ideal classes of order 3 in the class group of $\mathbb{Q}(\sqrt{-D})$. It is conjectured that the number of solutions, and hence the number of such ideal classes, should be $O(D^\varepsilon)$ for any $\varepsilon > 0$, but to date we cannot reduce the exponent below $1/2$. This question is related to the problem of giving an upper bound for the number of rational elliptic curves with given conductor.

7. It is conjectured that any irreducible polynomial $f(X) \in \mathbb{Z}[X]$ which satisfies the obvious congruence conditions should assume infinitely many square-free values. This has been established only for polynomials f of degree at most 3. What is required for further progress is a good bound for the number of solutions of the equation

$$f(x) = y^2 z, \quad 1 \leqslant x \leqslant N, \quad y \geqslant N.$$

These examples demonstrate that the general problem under consideration underlies a very diverse range of questions in number theory. Although many of the above examples involve inhomogeneous equations, we shall begin by considering only the case in which F is a form. Later on we shall see how the inhomogeneous case can be handled in an analogous way to the homogeneous one. We therefore state formally the following assumptions for future reference.

Convention *Unless stated explicitly otherwise, we shall write* **x** *for the vector* $(x_1, \ldots, x_n)$ *and assume that* $F(\mathbf{x}) \in \mathbb{Z}[\mathbf{x}]$ *is an absolutely irreducible form of degree* d.

1.3 The heuristic bounds

It will be convenient to define

$$N^{(0)}(B) = N^{(0)}(F; B) = \#\{\mathbf{x} \in \mathbb{Z}^n : F(\mathbf{x}) = 0, \ \max |x_i| \leqslant B\}.$$

Recall that F has total degree d. Then for the vectors **x** under consideration, the values $F(\mathbf{x})$ will all be of order B^d, and indeed a positive proportion of them will have exact order B^d. Thus the 'probability' that a randomly chosen value of $F(\mathbf{x})$ should vanish might be expected to be of order B^{-d}. Since the number of vectors **x** to be considered has order B^n, this heuristic argument leads one to expect that $N^{(0)}(B)$ is of exact order B^{n-d}.

Clearly there are many things wrong with this argument, not least the fact that when $n < d$ the solution $\mathbf{x} = \mathbf{0}$ shows that $N^{(0)}(B) \not\to 0$. However we can safely summarize things the following way.

Heuristic expectation *When $n \geqslant d$ we have*

$$B^{n-d} \ll N^{(0)}(B) \ll B^{n-d}$$

unless there is a reason why not!

Certainly local (congruence) conditions will often provide a reason why $N^{(0)}(B)$ counts only the solution $\mathbf{x} = \mathbf{0}$. However, in support of the above heuristic argument we have the following very general result of Birch.

Theorem 1 *Suppose $F(\mathbf{x})$ is a non-singular form of degree d in $n > 2^d(d-1)$ variables. Then there is a constant $c_F > 0$ such that*

$$N^{(0)}(B) \sim c_F B^{n-d},$$

providing that $F(\mathbf{x}) = 0$ has non-trivial solutions in $\mathbb{R}$ and each p-adic field $\mathbb{Q}_p$.

Since forms are in general non-singular, Birch's result answers our question completely for typical forms with $n > 2^d(d-1)$. It would be of considerable interest to reduce the lower bound for n, but except for $d \leqslant 3$ this has not been done. In view of Birch's theorem our interest will be centred on the case in which n is small compared with d.

As has been mentioned there are many cases in which $N^{(0)}(B)$ is not of order B^{n-d}. A good illustration is provided by the diagonal cubic equation

$$x_1^3 + x_2^3 + x_3^3 + x_4^3 = 0.$$

Here there are 'trivial' solutions of the type $(a, -a, b, -b)$ which already contribute $\gg B^2$ to $N^{(0)}(B)$.

If we use the counting function $N^{(0)}(B)$ we see that a single non-zero solution $F(\mathbf{x}_0) = 0$ will produce $\ll B$ scalar multiples, so that $N^{(0)}(B) \gg B$. This behaviour often masks the contribution of other solutions. Thus it is usually convenient to count only primitive solutions, where a non-zero vector $(x_1, \ldots, x_n)$ is said to be primitive if h.c.f.$(x_1, \ldots, x_n) = 1$. Indeed since the vector $-\mathbf{x}_0$ will be a solution of $F(\mathbf{x}) = 0$ if and only if $\mathbf{x}_0$ is also a solution, it is natural to define

$$N(B) = N(F; B) =$$
$$\frac{1}{2} \#\{\mathbf{x} \in \mathbb{Z}^n \; : \; F(\mathbf{x}) = 0, \; \text{h.c.f.}(x_1, \ldots, x_n) = 1, \; \max |x_i| \leqslant B\}. \quad (1.2)$$

1.4 Curves

When F is homogeneous and $n = 3$ the equation $F(x_1, x_2, x_3) = 0$ describes a projective curve in $\mathbb{P}^2$. In this situation a great deal is known. Such a curve has a genus g which is an integer in the range $0 \leqslant g \leqslant (d-1)(d-2)/2$. The generic curve of degree d will have $g = (d-1)(d-2)/2$.

When $g = 0$ the curve either has no rational points (as for example, when $F(\mathbf{x}) = x_1^2 + x_2^2 + x_3^2$) or it can be parameterized by rational functions. Such a parameterization allows us to estimate $N(B)$. For example, when $F(\mathbf{x}) = x_1^2 + x_2^2 - x_3^2$ the solutions take the form

$$(x_1, x_2, x_3) = \left(a(b^2 - c^2),\ 2abc,\ a(b^2 + c^2)\right) \text{ or } \left(2abc,\ a(b^2 - c^2),\ a(b^2 + c^2)\right).$$

It is then easy to see that $N(B)$ is precisely of order B^{n-d}, since $n - d = 1$ in this case.

On the other hand, a second example with genus $g = 0$ is provided by the cubic curve $x_1^2 x_2 - x_3^3 = 0$. In this case the solutions are proportional to $(a^3, b^3, a^2 b)$. One therefore sees that $N(B)$ is precisely of order $B^{2/3}$. In this example we have $n - d = 0 < 2/3$.

We now turn to curves of genus 1. Either such a curve has no rational points or it is an elliptic curve. In the latter case the set of rational points can be given an abelian group structure, and the Mordell–Weil Theorem tells us that the group has finite rank r, say. Moreover, it can be shown using Néron's theory of heights, that

$$N(B) \sim c_F (\log B)^{r/2} \tag{1.3}$$

where c_F is a non-zero constant. A cubic curve with a rational point is an elliptic curve, and in this case $n - d = 0$ so that $N(B)$ grows faster than B^{n-d} as soon as $r \geqslant 1$.

Finally we consider curves of genus $g \geqslant 2$. Here the celebrated theorem of Faltings [8] shows that there are finitely many rational points, so that

$$N(B) \ll_F 1. \tag{1.4}$$

1.5 Surfaces

We have already seen the example

$$x_1^3 + x_2^3 + x_3^3 + x_4^3 = 0 \tag{1.5}$$

in which the 'trivial' solutions already contribute $\gg B^2$ to $N^{(0)}(B)$. These trivial solutions satisfy the conditions $x_1 + x_2 = x_3 + x_4 = 0$, or $x_1 + x_3 = x_2 + x_4 = 0$, or $x_1 + x_4 = x_2 + x_3 = 0$. In each case the trivial solutions are those that lie on certain lines in $\mathbb{P}^3$. These lines lie in the surface (1.5), since the equations $x_1 + x_2 = x_3 + x_4 = 0$, for example, imply $x_1^3 + x_2^3 + x_3^3 + x_4^3 = 0$.

In general, we shall *define* a trivial solution to $F(x_1, x_2, x_3, x_4) = 0$ to be one that lies on a line which is contained in the corresponding surface.

Moreover we shall define $N_1(B)$ to be the counting function analogous to (1.2), but in which only non-trivial solutions are counted. Thus in the case $d = 3$, Manin's conjecture predicts the behaviour of $N_1(B)$.

In the example (1.5) we know a complete parametric solution, due to Euler, giving all the rational points as

$$
\begin{aligned}
x_1 &= (3b - a)(a^2 + 3b^2)c + c^4, \\
x_2 &= (3b + a)(a^2 + 3b^2)c - c^4, \\
x_3 &= (a^2 + 3b^2)^2 - (3b + a)c^3, \\
x_4 &= -(a^2 + 3b^2)^2 - (3b - a)c^3.
\end{aligned}
\tag{1.6}
$$

Although this produces all the rational points, it is unfortunately the case that the values of $x_1, \ldots, x_4$ may be integers with a large common factor, even when h.c.f.$(a, b, c) = 1$. In the absence of any good way to control such a common factor, Euler's formula is rather little use in producing an upper bound for $N_1(B)$. Indeed if one wishes to produce a lower bound, the obvious procedure is to use integral values $a, b, c \ll B^{1/4}$. In this way one can at best show that $N_1(B) \gg B^{3/4}$, while other methods yield lower bounds $N_1(B) \gg B$ and better. Thus even a complete parameterization of the solutions does not solve our problem.

A second instructive example is provided by the equation

$$
x_1^4 + x_2^4 = x_3^4 + x_4^4.
\tag{1.7}
$$

Here there is a family of non-trivial solutions given (also by Euler) as

$$
\begin{aligned}
x_1 &= a^7 + a^5 b^2 - 2a^3 b^4 + 3a^2 b^5 + ab^6, \\
x_2 &= a^6 b - 3a^5 b^2 - 2a^3 b^4 + a^2 b^5 + b^7, \\
x_3 &= a^7 + a^5 b^2 - 2a^3 b^4 - 3a^2 b^5 + ab^6, \\
x_4 &= a^6 b + 3a^5 b^2 - 2a^3 b^4 + a^2 b^5 + b^7.
\end{aligned}
\tag{1.8}
$$

Not all solutions have this form, but these suffice on taking $a, b \ll B^{1/7}$ to show that $N_1(B) \gg B^{2/7}$. (There is the primitivity condition on $\mathbf{x}$ to be dealt with, but this can be satisfactorily handled.) As a, b run over all possible values, the corresponding vectors $\mathbf{x}$ run over a curve in the surface (1.7). Indeed, since the curve is parameterized, it is a curve of genus zero. In this example therefore we see that a surface may contain a large number of points by virtue of there being a genus zero curve lying in the surface.

A related example is the Euler surface

$$
x_1^4 + x_2^4 + x_3^4 = x_4^4.
$$

Here it was shown by Elkies [4] that there is a genus 1 curve of positive rank lying in the surface. In view of Néron's result (1.3) this shows that $N_1(B) \gg (\log B)^{1/2}$. Thus we see that surfaces may contain infinitely many points when there is a curve Γ of genus 1 on the surface, such that Γ itself has infinitely many points.

In general we expect that this sort of behaviour is essentially all that can occur. The following is a consequence of a conjecture of Lang.

Conjecture 1 *A surface of general type contains only finitely many curves of genus zero or one, and contains only finitely many rational points not on one of these curves.*

The definition of "general type" is somewhat technical. However we note that a non-singular surface in $\mathbb{P}^3$ will be of general type as soon as $d \geqslant 5$.

1.6 Higher dimensions

For varieties of higher dimension there are analogous phenomena, and we have the following conjecture, which is again a consequence of Lang's conjecture.

Conjecture 2 *On a variety of general type all rational points belong to one of a finite number of proper subvarieties.*

2 Second lecture. A survey of results

In the remainder of these notes we shall allow all the constants implied by the $\ll$ and $O(\ldots)$ notations to depend on the degree d of the form F and on the number n of variables. However, where there is any further dependence on F we shall say so explicitly.

In this lecture, where results are formally stated as theorems, they will either be proved in full in the lectures that follow, or may be found in the author's paper [15].

2.1 Early approaches

Until recently there have been few general results giving bounds for $N(F; B)$. Perhaps the first, historically, is due to Cohen, in the appendix to the lecturer's paper [11], where it is shown that

$$N(F; B) \ll_{\varepsilon, F} B^{n-3/2+\varepsilon}$$

for any $\varepsilon > 0$, as soon as $d \geqslant 2$. The proof uses the large sieve, and information of the behaviour of F modulo many different primes.

A second approach uses exponential sums to a fixed modulus, the latter being chosen to have size a suitable power of B. To work effectively the method requires F to be non-singular. One can then show (Heath-Brown [12]) that

$$N(F; B) \ll_{\varepsilon, F} B^{n-2+2/(n+1)+\varepsilon}$$

for $d \geqslant 2$, and

$$N(F; B) \ll_{\varepsilon, F} B^{n-3+15/(n+5)+\varepsilon} \tag{2.1}$$

for $d \geqslant 3$, again for any $\varepsilon > 0$. This latter result yields the estimate

$$N(F; B) \ll_{\varepsilon, F} B^{n-2+\varepsilon}$$

for non-singular F of degree $d \geqslant 2$, as soon as $n \geqslant 10$.

Hooley [17] uses a sieve method rather different from Cohen's, which can be coupled with estimates for multi-dimensional exponential sums. Although no general results have been worked out, the method is quite efficient in those special cases for which it has been used. Thus for the equations (1.1), Hooley shows [18], [19] and [21] that one has

$$N_1(B) \ll_{\varepsilon, k} B^{5/3+\varepsilon}$$

when $k \geqslant 3$.

Other general methods depend on elementary differential geometry, as in Schmidt [29]. These techniques improve slightly on Cohen's result, and apply also to certain non-algebraic hypersurfaces.

2.2 The method of Bombieri and Pila

The most successful general method appears to be that introduced by Bombieri and Pila [2], and developed by the lecturer [15]. In their original work, Bombieri and Pila showed that if $f(x, y) \in \mathbb{Z}[x, y]$ is an absolutely irreducible polynomial of degree d, then

$$\#\{(x, y) \in \mathbb{Z}^2 : f(x, y) = 0, \ |x|, |y| \leqslant B\} \ll_{\varepsilon} B^{1/d+\varepsilon}. \tag{2.2}$$

Indeed their result was slightly more precise than this. One very important feature of this result is that it is completely uniform with respect to f.

The estimate (2.2) is essentially best possible, as the example $f(x, y) = x^d - y$ shows. Here there are $\ll B^{1/d}$ solutions $x = m$, $y = m^d$ with $m \ll B^{1/d}$.

Pila went on [28] to apply (2.2) to our general setting, and showed that

$$N(B) \ll_{\varepsilon} B^{n-2+1/d+\varepsilon}. \tag{2.3}$$

In the case of quadratic forms one can do better, and an elementary argument shows that

$$N(B) \ll_{\varepsilon} B^{n-2+\varepsilon}$$

if $d = 2$, see [15, Theorem 2]. However it is an interesting open question whether one can extend this to higher degree forms.

Conjecture 3 *For given $d \geqslant 3$ and $n \geqslant 3$ we have*

$$N(F; B) \ll_{\varepsilon} B^{n-2+\varepsilon}.$$

It would even be interesting to know whether this could be achieved with a possible dependence on F in the implied constant. The bound (2.1) achieves this for non-singular forms F, when $n \geqslant 10$.

The arguments of Bombieri and Pila [2], and of Pila [28], were essentially affine in nature, and only used properties of $\mathbb{A}^2$. This left open the question of whether there might be a natural extension to higher dimensions. In particular Pila's work [28] only used the bound (2.2), applying it to hyperplane sections of the variety $F = 0$.

2.3 Projective curves

The lecturer's work [15] extended the method of Bombieri and Pila to projective hypersurfaces of arbitrary dimension. We shall begin by describing the result obtained for curves.

Theorem 2 *Let $F(x_1, x_2, x_3) \in \mathbb{Z}[x_1, x_2, x_3]$ be an absolutely irreducible form of degree d, and let $\varepsilon > 0$. Then*

$$N(F; B) \ll_\varepsilon B^{2/d + \varepsilon}. \tag{2.4}$$

At first sight this is uninteresting, since it is clearly surpassed by the results of Néron (1.3) and Faltings (1.4). The difference however lies in the fact that (2.4) is uniform in F. If one tries to adapt the proof of (1.3), say, to investigate uniformity in F one finds that the rank of the curve comes into play. At present we have insufficient information about the size of the rank of elliptic curves to produce unconditional bounds, so the approach fails. None the less the lecturer has shown [14] that one has

$$N(F; B) \ll_\varepsilon B^\varepsilon$$

for non-singular cubic curves, under the assumption of the Birch–Swinnerton-Dyer conjecture and the Riemann Hypothesis for L-functions of elliptic curves. The second comment that must be made in relation to (2.4) is that it applies to curves of genus zero as well as to curves of genus one or more. If one looks at the example $F(\mathbf{x}) = x_1^d - x_2^{d-1} x_3$, then one sees that there are solutions $(m^{d-1}n, m^d, n^d)$, and these suffice to show that $N(B) \gg B^{2/d}$. It follows that (2.4) is essentially best possible. Moreover the exponent $2/d$ is clearly a considerable improvement on the value $1 + 1/d$ which the bound (2.3) would produce. None the less, the fact remains that the proof of Theorem 2 works only with the degree d, and fails to distinguish the genus of the curve. This is a serious defect in the approach.

In fact we need not require F to be absolutely irreducible in Theorem 2. Indeed for forms which are irreducible over $\mathbb{Q}$ but reducible over $\overline{\mathbb{Q}}$ we have the following stronger estimate.

Theorem 3 *Let $F(x_1, x_2, x_3) \in \overline{\mathbb{Q}}[x_1, x_2, x_3]$ be an absolutely irreducible form of degree d, but not a multiple of a rational form. Then $N(F; B) \leqslant d^2$.*

Moreover, if $F(x_1, x_2, x_3) \in \mathbb{Z}[x_1, x_2, x_3]$ *is a form of degree* d *which is irreducible over* $\mathbb{Q}$ *but not absolutely irreducible, then* $N(F; B) \leqslant d^2$.

To prove the first assertion one writes F as a linear combination $\sum \lambda_i F_i$ of rational forms F_i, with linearly independent λ_i. Some F_i is not a multiple of F, but all rational zeros of F must satisfy $F = F_i = 0$. The result then follows by Bézout's Theorem. The second statement clearly follows from the first, on splitting F into its irreducible factors over $\overline{\mathbb{Q}}$.

We can also estimate the number of points on curves in $\mathbb{P}^3$.

Theorem 4 *Let* C *be an irreducible curve in* $\mathbb{P}^3$, *of degree* d, *not necessarily defined over the rationals. Then* C *has* $O_\varepsilon(B^{2/d+\varepsilon})$ *primitive points* $\mathbf{x} \in \mathbb{Z}^4$ *in the cube* $\max |x_i| \leqslant B$.

This can be established by projecting C onto a suitable plane, and counting the points on the resulting plane curve. In general such a projection may have degree less than d. However, the generic projection has degree equal to d, so that the result follows providing that one chooses the projection map with suitable care.

As with the result of Bombieri and Pila, all the above bounds are completely independent of F. The key to this is the following result, in which we write $\|F\|$ for the height of the form F, defined as the maximum modulus of the coefficients of F.

Theorem 5 *Suppose that* $F(x_1, x_2, x_3) \in \mathbb{Z}[\mathbf{x}]$ *is a non-zero form of degree* d, *and that the coefficients of* F *have no common factor. Then either* $N(F; B) \leqslant d^2$ *or* $\|F\| \ll B^{d(d+1)(d+2)/2}$.

Thus, if one has a bound of the shape

$$N(F; B) \ll_\varepsilon B^{\theta+\varepsilon}\|F\|^\varepsilon,$$

valid for any $\varepsilon > 0$, then one can deduce that either $N(F; B) \ll 1$ or

$$N(F; B) \ll_\varepsilon B^{\theta+\varepsilon}B^{d(d+1)(d+2)\varepsilon/2}.$$

On re-defining ε we see in either case that

$$N(F; B) \ll_\varepsilon B^{\theta+\varepsilon}.$$

Thus the dependence on $\|F\|$ miraculously disappears!

The results described here should be compared with those in the work of Elkies [5]. The emphasis in [5] is on algorithms for searching for rational points. Elkies shows in [5, Theorem 3] that one can find the rational points of height at most B on a curve C of degree d, in time $O_{C,\varepsilon}(B^{2/d+\varepsilon})$. Thus in particular there are $O_{C,\varepsilon}(B^{2/d+\varepsilon})$ points to be found. Uniformity in C is not considered, but it seems quite plausible that the methods may yield a good dependence on the height of C, or even complete independence as in the theorems quoted above. The techniques used in the two papers show interesting similarities, although the precise relationship remains unclear. In fact Elkies also looks at surfaces and varieties of higher dimension, and gives a heuristic argument that leads to the same exponent $3/\sqrt{d}$ as occurs in Theorems 7 and 11 below.

2.4 Surfaces

In the previous lecture it was explained that the natural counting function for surfaces should exclude trivial solutions, and the function $N_1(B)$ was introduced. If there is a line L, defined over $\mathbb{Q}$, and lying in the surface $F(\mathbf{x}) = 0$, then points on L will contribute $\gg_F B^2$ to $N(F; B)$.

In analogy with Conjecture 3 we may now expect the following.

Conjecture 4 *Let* $F(x_1, x_2, x_3, x_4) \in \mathbb{Z}[x_1, x_2, x_3, x_4]$ *be an absolutely irreducible form of large degree* d, *and let* $\varepsilon > 0$. *Then*

$$N_1(F; B) \ll_\varepsilon B^{1+\varepsilon}.$$

This would be best possible, as the example

$$x_1^d + x_2^d - x_2^{d-2} x_3 x_4 = 0$$

shows. This surface is absolutely irreducible, and contains no lines other than those in the planes $x_2 = 0$, $x_3 = 0$ and $x_4 = 0$. However there are rational points $(0, ab, a^2, b^2)$, which yield $N_1(B) \gg B$.

That any points on lines in the surface will dominate the function $N(B)$ is shown by the following result.

Theorem 6 *For an absolutely irreducible form* $F(\mathbf{x}) \in \mathbb{Z}[x_1, \ldots, x_4]$ *of degree* 3 *or more, we have*

$$N_1(F; B) \ll_\varepsilon B^{52/27+\varepsilon}.$$

Moreover we can improve substantially on this for large values of d, as follows.

Theorem 7 *Let* $F(\mathbf{x}) \in \mathbb{Z}[x_1, \ldots, x_4]$ *be an absolutely irreducible form of degree* d. *Then we have*

$$N_1(F; B) \ll_\varepsilon B^{1+3/\sqrt{d}+\varepsilon}.$$

Surfaces of the type $G(x_1, x_2) = G(x_3, x_4)$, where G is a binary form, have been investigated fairly extensively in the past, although success has been limited to the cases in which $d = 3$ (Hooley [16], [22]), or $d = 4$ and G has the shape $ax^4 + bx^2y^2 + cy^4$ (Hooley [20]), or G is diagonal (Bennet, Dummigan and Wooley [1]). In the first two cases the methods save only a power of $\log B$ relative to B^2.

The above mentioned works were designed to show that almost all integers represented by G have essentially only one representation. We can now prove this for arbitrary irreducible forms. To formulate this precisely, we define an automorphism of the binary form G to be an invertible 2×2 matrix M, such that $G(M\mathbf{x}) = G(\mathbf{x})$ identically in $\mathbf{x}$. We then say that integral solutions of $G(\mathbf{x}) = n$ are equivalent if and only if they are related by such an automorphism with a rational matrix M. (One slightly strange consequence of this definition is that when $d = 1$ or $d = 2$ all non-zero integer solutions of $G(\mathbf{x}) = n$ are equivalent.)

We then have the following result.

Theorem 8 *Let $G(x, y) \in \mathbb{Z}[x, y]$ be a binary form of degree $d \geqslant 3$, irreducible over $\mathbb{Q}$. Then G has $O_d(1)$ automorphisms. Moreover the number of positive integers $n \leqslant X$ represented by the form G is of exact order $X^{2/d}$, providing that G assumes at least one positive value. Of these integers n there are $O_{\varepsilon,G}\left(X^{52/(27d-2)+\varepsilon}\right)$ for which there are two or more inequivalent integral representations.*

The statement that the number of representable integers is of exact order $X^{2/d}$ is a classical result of Erdős and Mahler [7], dating from 1938.

Although Theorem 6 shows that points on lines may predominate, it does not automatically verify Conjecture 3 for surfaces, since there may be infinitely many lines. However this possibility can be handled successfully, and we have the following result.

Theorem 9 *Let $F(\mathbf{x}) \in \mathbb{Z}[x_1, \ldots, x_4]$ be an absolutely irreducible form of degree $d \geqslant 2$. Then we have*

$$N(F; B) \ll_{\varepsilon} B^{2+\varepsilon}.$$

When F is non-singular we can do better than Theorem 6, and we have the following results.

Theorem 10 *Let $F(\mathbf{x}) \in \mathbb{Z}[x_1, \ldots, x_4]$ be a non-singular form of degree $d \geqslant 2$. Then we have*

$$N_1(F; B) \ll_{\varepsilon} B^{4/3+16/(9d)+\varepsilon}. \tag{2.5}$$

Theorem 11 *Let $F(\mathbf{x}) \in \mathbb{Z}[x_1, \ldots, x_4]$ be a non-singular form of degree $d \geqslant 2$. Then we have*

$$N_1(F; B) \ll_{\varepsilon} B^{1+\varepsilon} + B^{3/\sqrt{d}+2/(d-1)+\varepsilon}. \tag{2.6}$$

In particular

$$N_1(F; B) \ll_{\varepsilon} B^{1+\varepsilon}, \tag{2.7}$$

when $d \geqslant 13$. Let $N_2(F; B)$ be the number of points counted by $N(F; B)$, but not lying on any curve of degree $\leqslant d - 2$ contained in the surface. Then

$$N_2(F; B) \ll_{\varepsilon} B^{3/\sqrt{d}+2/(d-1)+\varepsilon}. \tag{2.8}$$

Let $N_3(F; B)$ be the number of points counted by $N(F; B)$, but not lying on any genus zero curve of degree $\leqslant d - 2$ contained in the surface. Then

$$N_3(F; B) \ll_{\varepsilon,F} B^{3/\sqrt{d}+2/(d-1)+\varepsilon}. \tag{2.9}$$

In particular we see that Conjecture 4 holds for $d \geqslant 13$, providing that F is non-singular. We note that the exponent in Theorem 10 is better for $2 \leqslant d \leqslant 5$, but that otherwise one should use Theorem 11.

It is plain that curves of low degree lying in the surface $F = 0$ are a potential source for a large contribution to $N(F; B)$. Thus the following geometric result, due to Colliot-Thélène, is of great significance.

Theorem 12 [15, Theorem 12] *Let $F(\mathbf{x}) = 0$ be a non-singular surface in $\mathbb{P}^3$, of degree d. Then for every degree $\delta \leqslant d - 2$ there is a constant $N(\delta, d)$, independent of F, such that the surface $F(\mathbf{x}) = 0$ contains at most $N(\delta, d)$ irreducible curves of degree δ.*

When $d = 3$ we have the familiar fact that a non-singular cubic surface has 27 lines. We can therefore take $N(1, 3) = 27$.

We now see from the estimate (2.9) that, with very few exceptions, the rational points on a non-singular surface of large degree are restricted to a finite number of curves of genus zero. This may be compared with the assertion of Conjecture 1.

The special diagonal surfaces

$$F(\mathbf{x}) = x_1^d + x_2^d - x_3^d - x_4^d = 0 \tag{2.10}$$

have received a great deal of attention, and it has been shown that

$$N_1(B) \ll B^{4/3+\varepsilon} \quad (d = 3)$$

(Heath-Brown [13]),

$$N_1(B) \ll B^{5/3+\varepsilon} \quad (4 \leqslant d \leqslant 7) \tag{2.11}$$

(Hooley [19] and [21]), and

$$N_1(B) \ll B^{3/2+1/(d-1)+\varepsilon} \quad (d \geqslant 8)$$

(Skinner and Wooley [30]). Theorem 11 supersedes these as soon as $d \geqslant 6$. However Browning [3] has recently shown that (2.6) may be replaced by

$$N_1(F; B) \ll_\varepsilon B^{2/3+\varepsilon} + B^{3/\sqrt{d}+2/(d-1)+\varepsilon}$$

for these particular surfaces. We shall improve this further as follows.

Theorem 13 *When $d \geqslant 8$ the surface (2.10) contains no genus zero curves other than the lines. Hence*

$$N_1(F; B) \ll_\varepsilon B^{3/\sqrt{d}+2/(d-1)+\varepsilon}$$

for any $d \geqslant 2$.

2.5 A general result

We may now state the key result underlying most of the estimates described in this lecture. It applies to projective hypersurfaces of arbitrary dimension.

Theorem 14 *Let $F(x_1, \ldots, x_n) \in \mathbb{Z}[\mathbf{x}]$ be an absolutely irreducible form of degree d, and let $\varepsilon > 0$ and $B \geqslant 1$ be given. Then we can find $D = D(n, d, \varepsilon)$ and an integer k with*

$$k \ll_{n,d,\varepsilon} B^{(n-1)d^{-1/(n-2)}+\varepsilon} (\log \|F\|)^{2n-3},$$

as follows. There are forms $F_1(\mathbf{x}), \ldots, F_k(\mathbf{x}) \in \mathbb{Z}[x_1, \ldots, x_n]$, coprime to $F(\mathbf{x})$ and with degrees at most D, such that every point $\mathbf{x}$ counted by $N(F; B)$ is a zero of some form $F_j(\mathbf{x})$.

One should note that it is crucial for the degrees of the forms F_j to be suitably bounded, since one can easily construct a form $F_1(\mathbf{x})$, with degree dependent on B, which vanishes at every integer vector in the cube $\max |x_i| \leqslant B$.

In the case $n = 3$, Theorem 14 shows that every point counted by $N(F; B)$ satisfies $F(\mathbf{x}) = F_j(\mathbf{x}) = 0$ for some $j \ll_\varepsilon B^{2/d+\varepsilon} (\log \|F\|)^3$. Bézout's Theorem shows that there are at most dD points for each j, so that $N(F; B) \ll_\varepsilon B^{2/d+\varepsilon} (\log \|F\|)^3$. Theorem 2 then follows via an application of Theorem 5.

For the case $n = 4$ we see in the same way that the relevant points will lie on $O_\varepsilon (B^{3/\sqrt{d}+\varepsilon} (\log \|F\|)^5)$ curves in the surface $F = 0$, each curve having degree at most dD. We may apply Theorem 4 to estimate the number of points on such a curve, but it is useful to have further information on the possible degrees of such curves. This is provided by Theorem 12 when the form F is non-singular, but otherwise we merely use the fact that the degree of the curve will be at least 2 for points counted by $N_1(F; B)$. In this way we may establish Theorems 7 and 11, after treating the factor $(\log \|F\|)^5$ through a version of the process employed for Theorem 5.

2.6 Affine problems

Although our main emphasis has been on integer zeros of forms, one can successfully tackle problems involving general (inhomogeneous) polynomials. For this section we therefore suppose that $F(x_1, \ldots, x_n) \in \mathbb{Z}[\mathbf{x}]$ is an absolutely irreducible polynomial of total degree d, and we consider

$$N(F; B_1, \ldots, B_n) = N(F; \mathbf{B}) =$$

$$\#\{\mathbf{x} \in \mathbb{Z} : F(\mathbf{x}) = 0, \ |x_i| \leqslant B_i, \ (1 \leqslant i \leqslant n)\} \quad (2.12)$$

where $B_i \geqslant 1$ for $1 \leqslant i \leqslant n$. Thus the Bombieri–Pila result (2.2) shows that $N(F; B, B) \ll_\varepsilon B^{1/d+\varepsilon}$. In applications it can be very useful to allow the B_i

to have varying sizes. Indeed one can formulate the homogeneous problem with general boxes rather than cubes, and prove an extension of Theorem 14. In our case we shall see that the following analogue of Theorem 14 holds.

Theorem 15 *Let $F(x_1, \ldots, x_n) \in \mathbb{Z}[\mathbf{x}]$ be an absolutely irreducible polynomial of degree d, and let $\varepsilon > 0$ and $B_1, \ldots, B_n \geqslant 1$ be given. Define*

$$T = \max \left\{ \prod_{i=1}^n B_i^{e_i} \right\},$$

where the maximum is taken over all integer n-tuples $(e_1, \ldots, e_n)$ for which the corresponding monomial

$$x_1^{e_1} \ldots x_n^{e_n}$$

occurs in $F(\mathbf{x})$ with non-zero coefficient.

Then we can find $D = D(n, d, \varepsilon)$ and an integer k with

$$k \ll_{n,d,\varepsilon} T^\varepsilon \exp \left\{ (n-1) \left(\frac{\prod \log B_i}{\log T} \right)^{1/(n-1)} \right\} (\log \|F\|)^{2n-3},$$

as follows. There are polynomials $F_1(\mathbf{x}), \ldots, F_k(\mathbf{x}) \in \mathbb{Z}[x_1', \ldots, x_n]$, coprime to $F(\mathbf{x})$ and with degrees at most D, such that every point $\mathbf{x}$ counted by $N(F; \mathbf{B})$ is a zero of some polynomial $F_j(\mathbf{x})$.

3 Third lecture. Proof of Theorem 14

3.1 Singular points

In proving Theorem 14 we shall begin by considering singular points. A singular point of $F(\mathbf{x}) = 0$ satisfies

$$\frac{\partial F(\mathbf{x})}{\partial x_i} = 0, \quad (1 \leqslant i \leqslant n).$$

Not all the forms $\partial F/\partial x_i$ can be identically zero, since F is absolutely irreducible. Moreover, if one of the partial derivatives is non-zero it will have degree $d - 1$, so that it cannot be a multiple of F. Thus if we include a nonzero partial derivative amongst the forms F_j, all singular points will be taken care of.

Our proof of Theorem 14 will use an auxiliary prime p, and we shall also need to account for points which are singular modulo p. We set

$$S(F; B, p) = \left\{ \mathbf{x} \in \mathbb{Z}^n : F(\mathbf{x}) = 0, \ |x_i| \leqslant B, \ (1 \leqslant i \leqslant n), \ p \nmid \nabla F(\mathbf{x}) \right\},$$

and

$$S(F; B) = \left\{ \mathbf{x} \in \mathbb{Z}^n : F(\mathbf{x}) = 0, \ |x_i| \leqslant B, \ (1 \leqslant i \leqslant n), \ \nabla F(\mathbf{x}) \neq \mathbf{0} \right\}.$$

The following lemma then holds.

Lemma 1 *Let $B \geqslant 2$ and $r = \lceil \log(\|F\|B) \rceil$, and assume that*

$$P \geqslant \log^2(\|F\|B).$$

Then we can find primes $p_1 < \ldots < p_r$ in the range $P \ll p_i \ll P$, such that

$$S(F; B) = \bigcup_{i=1}^{r} S(F; B, p_i).$$

In fact this is the only place in the argument where a dependence on $\|F\|$ occurs.

For the proof we just pick the first r primes $p_i > AP$, with a suitable constant A. We will then have $P \ll p_i \ll P$ since $P \gg r^2$. For any $\mathbf{x} \in S(F; B)$ there will be some partial derivative $\partial F/\partial x_j$, say, which is non-zero. Using the bound

$$\frac{\partial F}{\partial x_j} \ll_n \|F\|B^{d-1},$$

we see that

$$\#\left\{ p > AP : p \left| \frac{\partial F}{\partial x_j} \right. \right\} \ll_{n,d} \frac{\log(\|F\|B)}{\log(AP)}.$$

Thus there are at most $r - 1$ such primes, if A is large enough. It follows that one of the primes p_i does not divide $\partial F/\partial x_j$, in which case $\mathbf{x} \in S(F; B, p_i)$, as required.

As a consequence of Lemma 1 it suffices to examine points which are non-singular modulo a fixed prime $p \gg \log^2(\|F\|B)$, providing that we allow an extra factor $\log(\|F\|B)$ in our final estimate for k.

3.2 The Implicit Function Theorem

We shall write

$$S(F; B, p) = \bigcup_{\mathbf{t}} S(\mathbf{t}),$$

where

$$S(\mathbf{t}) = \left\{ \mathbf{x} \in S(F; B, p) : \mathbf{x} \equiv \rho \mathbf{t} \pmod{p} \text{ for some } \rho \in \mathbb{Z} \right\},$$

and $\mathbf{t}$ runs over a set of projective representatives for the non-singular points of $F(\mathbf{t}) = 0$ over $\mathbb{F}_p$.

The proof of Theorem 14 will show that if p is sufficiently large compared with B, then all points $\mathbf{x} \in S(\mathbf{t})$ satisfy an equation $F(\mathbf{x}; \mathbf{t}) = 0$. The forms $F(\mathbf{x}; \mathbf{t})$ will turn out to have the properties described in Theorem 14, so that if we take k' to be the number of non-singular points of $F(\mathbf{t}) = 0$ over $\mathbb{F}_p$, we will have $k \ll 1 + k'(\log \|F\|B)$, according to the argument above. Indeed we will have $k' \ll p^{n-2} \ll P^{n-2}$, so that it will be enough to show that

$$P \gg B^{(n-1)(n-2)^{-1}d - 1/(n-2)} V^{\varepsilon} \log^2 \|F\| \tag{3.1}$$

suffices. Note in particular that (3.1) certainly ensures that $P \gg \log^2(\|F\|B)$, when B is large enough, so that Lemma 1 applies.

We shall now fix our attention on a particular value of $\mathbf{t}$. Without loss of generality we may take $t_1 - 1$, since we are working in projective space. If the partial derivatives

$$\frac{\partial F}{\partial x_i}(\mathbf{t}) \tag{3.2}$$

were to vanish for $2 \leqslant i \leqslant n$, then the first partial derivative must vanish too, since

$$0 = dF(\mathbf{t}) = \mathbf{t} \cdot \nabla F(\mathbf{t}).$$

However $\mathbf{t}$ was assumed to be non-singular, so there must be some non-vanishing partial derivative with $2 \leqslant i \leqslant n$. Without loss of generality we shall assume that in fact

$$\frac{\partial F}{\partial x_2}(\mathbf{t}) \neq 0. \tag{3.3}$$

Using Hensel's lemma, along with (3.3), we can lift $\mathbf{t}$ to a p-adic solution $\mathbf{u} \in \mathbb{Z}_p^n$ of $F(\mathbf{u}) = 0$ in which $u_1 = 1$. One can now show that the equation

$$F(1, u_2 + Y_2, u_3 + Y_3, \ldots, u_n + Y_n) = 0$$

may be used to define Y_2 implicitly as a convergent p-adic power series in $Y_3, \ldots, Y_n$, providing that $p \mid Y_i$ for $2 \leqslant i \leqslant n$. This is, in effect, an application of the implicit function theorem, but we shall formulate it in terms of polynomials, as follows.

Lemma 2 *Let $F(\mathbf{x})$ and $\mathbf{u}$ be as above. Then for any integer $m \geqslant 1$ we can find $f_m(Y_3, Y_4, \ldots, Y_n) \in \mathbb{Z}_p[Y_3, \ldots, Y_n]$, such that if $F(\mathbf{v}) = 0$ for some $\mathbf{v} \in \mathbb{Z}_p^n$ with $v_1 = 1$ and $\mathbf{v} \equiv \mathbf{u} \pmod{p}$, then*

$$v_2 \equiv f_m(v_3, \ldots, v_n) \pmod{p^m}. \tag{3.4}$$

We shall prove Lemma 2 by induction on m. Write

$$\frac{\partial F}{\partial x_2}(\mathbf{u}) = \mu,$$

say, and let $f_1(Y_3, \ldots, Y_n) = u_2$ (constant), and

$$f_{m+1}(Y_3, \ldots, Y_n) = f_m(Y_3, \ldots, Y_n) - \mu^{-1} F(1, f_m(Y_3, \ldots, Y_n), Y_3, \ldots, Y_n),$$

for $m \geqslant 1$. The case $m = 1$ of Lemma 2 is then immediate. For the general case the induction hypothesis yields

$$v_2 \equiv f_m(v_3, \ldots, v_n) \pmod{p^m},$$

so that we may write

$$v_2 = f_m(v_3, \ldots, v_n) + \lambda p^m,$$

with $\lambda \in \mathbb{Z}_p$. Then

$$
\begin{aligned}
0 &= F(\mathbf{v}) \\
&\equiv F(1, f_m(v_3, \ldots, v_n), v_3, \ldots, v_n) \\
&\quad + \lambda p^m \frac{\partial F}{\partial x_2}(1, f_m(v_3, \ldots, v_n), v_3, \ldots, v_n) \pmod{p^{m+1}}. \quad (3.5)
\end{aligned}
$$

Moreover, the induction hypothesis (3.4) shows that

$$f_m(v_3, \ldots, v_n) \equiv u_2 \pmod{p},$$

since $\mathbf{v} \equiv \mathbf{u} \pmod{p}$. It follows that

$$\frac{\partial F}{\partial x_2}(1, f_m(v_3, \ldots, v_n), v_3, \ldots, v_n) \equiv \mu \pmod{p}.$$

We now see from (3.5) that

$$\lambda p^m \equiv -\mu^{-1} F(1, f_m(v_3, \ldots, v_n), v_3, \ldots, v_n) \pmod{p^{m+1}},$$

so that

$$v_2 \equiv f_{m+1}(v_3, \ldots, v_n) \pmod{p^{m+1}}.$$

This completes the induction.

3.3 Vanishing determinants of monomials

Clearly we can apply an invertible integral linear transformation to the form $F(\mathbf{x})$ so as to produce a form in which the coefficient of x_n^d is non-zero. Indeed we can find such a transformation in which the coefficients are all $O_{d,n}(1)$. Thus there is no loss of generality in assuming that the monomial x_n^d has non-zero coefficient in $F(\mathbf{x})$.

We now choose a large integer D and define the set

$$\mathcal{E} = \left\{ (e_1, \ldots, e_n) \in \mathbb{Z}^n \; : \; e_i \geqslant 0, \; (1 \leqslant i \leqslant n), \; e_n < d, \; \sum_{i=1}^{n} e_i = D \right\}. \quad (3.6)$$

We shall write

$$E = \#\mathcal{E} = \binom{D+n-1}{n-1} - \binom{D-d+n-1}{n-1}, \quad (3.7)$$

and we shall suppose for the moment that $E \leqslant \#S(\mathbf{t})$. Let $\mathbf{x}^{(1)}, \ldots, \mathbf{x}^{(E)}$ be distinct vectors in $S(\mathbf{t})$ and let

$$\Delta = \det\left(\mathbf{x}^{(i)\mathbf{e}}\right)_{1\leqslant i\leqslant E,\ \mathbf{e}\in\mathcal{E}}$$

where we write

$$w_1^{e_1}\dots w_n^{e_n} = \mathbf{w}^{\mathbf{e}}.$$

Thus Δ is an $E \times E$ determinant with rows corresponding to the different vectors $\mathbf{x}^{(i)}$ and columns corresponding to the various exponent n-tuples $\mathbf{e}$.

We proceed to show that Δ is divisible by a large power p^m of p. This will enable us to deduce that Δ vanishes. Since $\mathbf{x}^{(i)} \in S(\mathbf{t})$, we see that the reduction modulo p of $\mathbf{x}^{(i)}$ represents the same projective point as $\mathbf{t}$ does. It follows that $p \nmid x_1$ so that we may regard $x_1^{-1}\mathbf{x} = \mathbf{v}$, say, as a vector in $\mathbb{Z}_p^n$. We now have $v_1 = t_1 = 1$. Moreover we may lift $\mathbf{t}$ to a vector $(1, u_2, u_3, \dots, u_n) \in \mathbb{Z}_p^n$ on $F = 0$, as in Lemma 2, so that $v_i = u_i + y_i$ for $2 \leqslant i \leqslant n$, for suitable $y_i \in p\mathbb{Z}_p$. Lemma 2 now shows that

$$\Delta = \left(\prod_{1\leqslant i\leqslant E} x_1^{(i)}\right)^D \det\left(\mathbf{v}^{(i)\mathbf{e}}\right)_{1\leqslant i\leqslant E,\ \mathbf{e}\in\mathcal{E}} \equiv \left(\prod_{1\leqslant i\leqslant E} x_1^{(i)}\right)^D \Delta_0 \pmod{p^m},$$

where

$$\Delta_0 = \det(M_0), \qquad M_0 = \left(\mathbf{w}^{(i)\mathbf{e}}\right)_{1\leqslant i\leqslant E,\ \mathbf{e}\in\mathcal{E}},$$

with

$$w_1^{(i)} = 1, \qquad w_2^{(i)} = f_m\left(v_3^{(i)}, \dots, v_n^{(i)}\right),$$

and

$$w_j^{(i)} = v_j^{(i)} \qquad (3 \leqslant j \leqslant n).$$

We proceed to replace $v_j^{(i)}$ by $u_j + y_j^{(i)}$ for $3 \leqslant j \leqslant n$, so that we have $p \mid y_j^{(i)}$. It follows that

$$\mathbf{w}^{(i)\mathbf{e}} = w_1^{(i)e_1}\dots w_n^{(i)e_n} = g_{\mathbf{e}}\left(y_3^{(i)}, y_4^{(i)}, \dots, y_n^{(i)}\right)$$

for a suitable collection of polynomials $g_{\mathbf{e}}(Y_3, \dots, Y_n) \in \mathbb{Z}_p[Y_3, \dots, Y_n]$. We now choose an ordering $\prec$ on the vectors

$$\mathbf{f} = (f_3, \dots, f_n), \qquad (f_j \in \mathbb{Z},\ f_j \geqslant 0),$$

in such a way that $\mathbf{f} \prec \mathbf{f}'$ if $\sum f_j < \sum f_j'$. (When $n \geqslant 4$ this can be done in many different ways.) We then order the monomials $\mathbf{Y}^{\mathbf{f}}$ in the corresponding fashion.

We now perform column operations on M_0 using the following procedure. We take the 'smallest' monomial $\mathbf{Y}^{\mathbf{f}}$, say, occurring in any of the polynomials $g_{\mathbf{e}}$. If this monomial occurs in two or more such polynomials, we use the monomial for which the p-adic order of the coefficient is least. We then interchange columns so as to bring this term into the leading column, and proceed to subtract p-adic integer multiples of the new first column from any other columns which contain the monomial $\mathbf{Y}^{\mathbf{f}}$. Thus this monomial will now occur only in the first column. We then repeat the procedure with the remaining $n - 1$ columns, looking again for the 'smallest' monomial, placing it in the second column, and removing it from all later columns. Continuing in this manner we reach an expression

$$\Delta_0 = \det(M_1), \quad M_1 = \left(h_e \left(y_3^{(i)}, \ldots, y_n^{(i)} \right) \right)_{1 \leqslant i \leqslant E, \, 1 \leqslant e \leqslant E}$$

in which we have polynomials $h_e(\mathbf{Y}) \in \mathbb{Z}_p[\mathbf{Y}]$, with successively larger 'smallest' monomial terms.

There are

$$\binom{f + n - 3}{n - 3} = n(f),$$

say, monomials of total degree f. Hence if $e > n(0) + n(1) + \ldots + n(f - 1)$, the 'smallest' monomial in $h_e(\mathbf{Y})$ will have total degree at least f. We now recall that $p \, | \, y_j^{(i)}$ for $3 \leqslant j \leqslant n$ whence every element in the e-th column of M_1 will be divisible by p^f. Since

$$\sum_{i=0}^{f} n(i) = \binom{f + n - 2}{n - 2},$$

and

$$\sum_{i=0}^{f} i \, n(i) = (f + 1) \binom{f + n - 2}{n - 2} - \binom{f + n - 1}{n - 1},$$

it follows that if

$$\binom{f + n - 2}{n - 2} \leqslant E < \binom{(f + 1) + n - 2}{n - 2}, \tag{3.8}$$

then Δ_0 is divisible by

$$p^{n(1) + 2n(2) + \ldots + fn(f) + (f+1)(E - n(0) - n(1) - \ldots - n(f))} = p^{\nu},$$

say, where

$$\nu = (f + 1)E - \binom{f + n - 1}{n - 1}. \tag{3.9}$$

We therefore specify that the prime power p^m with which we work will have $m = \nu$. This leads to the following conclusion.

Lemma 3 *If E lies in the interval (3.8) and ν is as in (3.9), then*

$$\nu_p(\Delta) \geqslant \nu.$$

To show that Δ must in fact vanish we shall use information on its size. Each entry in Δ has modulus at most B^D, whence

$$|\Delta| \leqslant E^E B^{DE}.$$

Hence if we impose the condition

$$p^\nu > E^E B^{DE}, \tag{3.10}$$

we deduce from Lemma 3 that $\Delta = 0$.

3.4 Completion of the proof

We are now ready to construct the form F_j corresponding to our chosen value of $\mathbf{t}$. Recall that we have supposed that $\#S(\mathbf{t}) \geqslant E$, and that we took the points $\mathbf{x}^{(1)}, \ldots, \mathbf{x}^{(E)}$ to be distinct elements of $S(\mathbf{t})$. We now set $\#S(\mathbf{t}) = K$ and examine the matrix

$$M_2 = \left(\mathbf{x}^{(i)\mathbf{e}}\right)_{1 \leqslant i \leqslant K,\, \mathbf{e} \in \mathcal{E}}$$

where the vectors $\mathbf{x}^{(i)}$ now run over all elements of $S(\mathbf{t})$. From what we have proved it follows that M_2 has rank at most $E - 1$. This is trivial if $K \leqslant E - 1$, and otherwise the work of the previous section shows that every $E \times E$ minor vanishes. We therefore see that $M_2 \mathbf{c} = \mathbf{0}$ for some non-zero vector $\mathbf{c} \in \mathbb{Z}^E$. Hence if

$$F_j(\mathbf{x}) = \sum_{\mathbf{e} \in \mathcal{E}} c_{\mathbf{e}} \mathbf{x}^{\mathbf{e}}, \tag{3.11}$$

we will have a non-zero form, of degree D, which vanishes for every $\mathbf{x} \in S(\mathbf{t})$. Theorem 14 requires that $F(\mathbf{x})$ does not divide $F_j(\mathbf{x})$. However this is clear from our choice of the exponent set $\mathcal{E}$, since F contains a term in x_n^d, whereas F_j does not.

From (3.10) we see that it suffices to have $p \gg_D B^{DE/\nu}$, so that it will be enough to prove that

$$\lim_{D \to \infty} \frac{DE}{\nu} = \frac{n-1}{n-2} d^{-1/(n-2)}.$$

In view of (3.1) this will complete the proof of Theorem 14.

From (3.7) we have

$$E = \frac{d\, D^{n-2}}{(n-2)!} + O(D^{n-3}),$$

where the implied constant may depend on n and d. On the other hand, (3.8) implies that

$$E = \frac{f^{n-2}}{(n-2)!} + O(f^{n-3}).$$

We therefore deduce that

$$f = d^{1/(n-2)} D + O(1),$$

whence (3.9) yields

$$\nu = \frac{(n-2)f^{n-1}}{(n-1)!} + O(f^{n-2})$$

$$= d^{(n-1)/(n-2)}(n-2)\frac{D^{n-1}}{(n-1)!} + O(D^{n-2}). \tag{3.12}$$

These estimates then produce the required limiting behaviour for DE/ν, thereby completing the proof of Theorem 14.

We end this lecture by establishing Theorem 5. For convenience we set $M = (d+1)(d+2)/2$ and $N = d^2 + 1$. We then suppose that $F(\mathbf{x}) = 0$ has solutions $\mathbf{x}^{(1)}, \ldots, \mathbf{x}^{(N)} \in \mathbb{Z}^3$, with $|\mathbf{x}^{(i)}| \ll B$. We now consider a matrix C of size $N \times M$, in which the i-th row consists of the M possible monomials of degree d in the variables $x_1^{(i)}$, $x_2^{(i)}$, $x_3^{(i)}$. Let $\mathbf{f} \in \mathbb{Z}^M$ have entries which are the corresponding coefficients of F, so that $C\mathbf{f} = \mathbf{0}$. It follows that C has rank at most $M - 1$, since $\mathbf{f}$ is non-zero. We therefore see that $C\mathbf{g} = \mathbf{0}$ has a non-zero integer solution $\mathbf{g}$, which can be constructed from the various sub-determinants of C. It follows that $|\mathbf{g}| \ll_d B^{dM}$. Now take $G(\mathbf{x})$ to be the ternary form, of degree d, corresponding to the coefficient vector $\mathbf{g}$. By our construction, $G(\mathbf{x})$ and $F(\mathbf{x})$ have common zeros at each of the points $\mathbf{x}^{(i)}$ for $1 \leqslant i \leqslant d^2 + 1$. This will contradict Bézout's Theorem, unless F and G are proportional. In the latter case we may deduce that $\|F\| \ll_d \|G\| \ll_d B^{dM}$, since the coefficients of F have no common factor. This gives us the required conclusion.

4 Fourth lecture. Rational points on projective surfaces

This lecture will be devoted to the proofs of Theorems 6 and 8. We shall not present all the details, for which the reader should consult [15].

4.1 Theorem 6 – Plane sections

The principal tool for the proof of Theorem 6 is the following result from the geometry of numbers, for which see [15, Lemma 1, parts (iii) and (iv)].

Lemma 4 Let $\mathbf{x} \in \mathbb{Z}^n$ lie in the cube $|x_i| \leqslant B$. Then there is a primitive vector $\mathbf{z} \in \mathbb{Z}^n$, for which $\mathbf{x} \cdot \mathbf{z} = 0$, and such that $|\mathbf{z}| \ll B^{1/(n-1)}$.

Moreover, if $\mathbf{z} \in \mathbb{Z}^n$ is primitive, then there exist primitive vectors

$$\mathbf{b}^{(1)}, \ldots, \mathbf{b}^{(n-1)} \in \mathbb{Z}^n$$

such that

$$|\mathbf{z}| \ll \prod_{j=1}^{n-1} |\mathbf{b}^{(j)}| \ll |\mathbf{z}|. \tag{4.1}$$

These have the property that any vector $\mathbf{x} \in \mathbb{Z}^n$ *with* $\mathbf{x}\cdot\mathbf{z} = 0$ *may be written as a linear combination*

$$\mathbf{x} = \lambda_1 \mathbf{b}^{(1)} + \ldots + \lambda_{n-1} \mathbf{b}^{(n-1)} \qquad (4.2)$$

with $\lambda_1, \ldots, \lambda_{n-1} \in \mathbb{Z}$ *and*

$$\lambda_j \ll \frac{|\mathbf{x}|}{|\mathbf{b}^{(j)}|}, \qquad (1 \leqslant j \leqslant n-1).$$

It follows that the region $|\mathbf{x}| \leqslant X$ *contains* $O(X^{n-1}/|\mathbf{z}|)$ *integral vectors orthogonal to* $\mathbf{z}$, *providing that* $X \gg |\mathbf{z}|$.

For the last part we note that $|\mathbf{b}^{(j)}| \ll |\mathbf{z}| \ll X$, by (4.1), and hence that there are $O(X/|\mathbf{b}^{(j)}|)$ choices for each λ_j. The result then follows from a second application of (4.1).

Taking $n = 4$ we see that every relevant point on the surface $F(\mathbf{x}) = 0$ must lie on one of $O(B^{4/3})$ planes $\mathbf{x}\cdot\mathbf{y} = 0$ with $|\mathbf{y}| \ll B^{1/3}$. We proceed to count the number of points of the surface $F(\mathbf{x}) = 0$ which lie on a given plane. According to Lemma 4, each vector $\mathbf{y}$ determines a triple of vectors $\mathbf{b}^{(1)}, \mathbf{b}^{(2)}, \mathbf{b}^{(3)}$. We may then write $\mathbf{x}$ in the form (4.2) and substitute into the equation $F(\mathbf{x}) = 0$ to obtain a condition $G(\lambda_1, \lambda_2, \lambda_3) = 0$, say, in which G is an integral form of degree d, though not necessarily irreducible. If $\mathbf{x}$ is primitive then it is clear that $(\lambda_1, \lambda_2, \lambda_3) \in \mathbb{Z}^3$ is also primitive. Moreover, if we choose $|\mathbf{b}^{(1)}| \leqslant |\mathbf{b}^{(2)}| \leqslant |\mathbf{b}^{(3)}|$, then the condition $\max |x_i| \leqslant B$ implies $\max |\lambda_i| \leqslant cB/|\mathbf{b}^{(1)}|$ for some absolute constant c.

4.2 Theorem 6 – Curves of degree 3 or more

If H is an absolutely irreducible factor of G, then the solutions of $H(\lambda_1, \lambda_2, \lambda_3) = 0$ will correspond to points on an irreducible plane curve in the surface $F = 0$. If H has degree at least 3, then

$$N\big(H; cB/|\mathbf{b}^{(1)}|\big) \ll_\varepsilon \left(\frac{B}{|\mathbf{b}^{(1)}|}\right)^{2/3+\varepsilon}, \qquad (4.3)$$

by Theorem 2. It is important to notice that the implied constant is independent of H, and hence of the vectors $\mathbf{b}^{(j)}$ and $\mathbf{y}$. The estimate (4.3) allows us to bound the number of solutions of $G(\lambda_1, \lambda_2, \lambda_3) = 0$ arising from all factors H of G with degree 3 or more. We proceed to sum the bound (4.3) over those vectors $\mathbf{y}$ that arise. In order to do this we need to count how many vectors $\mathbf{y}$ can correspond to a given $\mathbf{b}^{(1)}$. To do this we apply the final part of Lemma 4, taking $\mathbf{z}$ to be $\mathbf{b}^{(1)}$. We then see that there are $O(B/|\mathbf{b}^{(1)}|)$ possible vectors $\mathbf{y}$ in the region $|\mathbf{y}| \ll B^{1/3}$ which are orthogonal to a given $\mathbf{b}^{(1)}$, so that the total contribution of the estimates (4.3), when we sum over $\mathbf{y}$, is

$$\ll_\varepsilon \sum \left(\frac{B}{|\mathbf{b}^{(1)}|} \right)^{5/3+\varepsilon}, \tag{4.4}$$

the sum being over the possible vectors $\mathbf{b}^{(1)}$. Since we took $|\mathbf{b}^{(1)}| \leqslant |\mathbf{b}^{(2)}| \leqslant |\mathbf{b}^{(3)}|$, and we also have

$$|\mathbf{y}| \ll \prod_{j=1}^{n-1} |\mathbf{b}^{(j)}| \ll |\mathbf{y}|$$

by (4.1), it follows that $|\mathbf{b}^{(1)}| \ll |\mathbf{y}|^{1/3} \ll B^{1/9}$. The sum (4.4) is therefore

$$\ll_\varepsilon B^{5/3+\varepsilon} \left(B^{1/9} \right)^{4-5/3-\varepsilon} \ll_\varepsilon B^{52/27+\varepsilon}.$$

This is satisfactory for Theorem 6.

4.3 Theorem 6 – Quadratic curves

It remains to handle the case in which $(\lambda_1, \lambda_2, \lambda_3)$ is a zero of a linear or quadratic factor H of G. Here we shall be brief. If H is linear, then the equations $\mathbf{x} \cdot \mathbf{y} = 0$ and $H(\lambda_1, \lambda_2, \lambda_3) = 0$ describe a line in the surface $F = 0$, so that the corresponding points $\mathbf{x}$ are not counted by $N_1(F; B)$. According to Harris [10, Proposition 18.10], the generic plane section of any irreducible hypersurface is itself irreducible. Thus the set of vectors $\mathbf{y}$ for which the corresponding form G is reducible must lie on a certain union of irreducible surfaces in $\mathbb{P}^3$ (or possibly indeed in some smaller algebraic set). One can show [15, §6] that the number and degrees of the components of this set may be bounded in terms of d, so that there are $O(Y^3)$ admissible vectors $\mathbf{y}$ with $|\mathbf{y}| \leqslant Y$. To obtain the estimate $O(Y^3)$ one may apply Pila's result (2.3) to components of degree 2 or more, and the trivial bound for linear components. In the case in which H is quadratic it can be shown that there are $O_\varepsilon(B^{1+\varepsilon}|\mathbf{y}|^{-1/3})$ solutions $(\lambda_1, \lambda_2, \lambda_3)$. Thus vectors $\mathbf{y}$ with $Y/2 < |\mathbf{y}| \leqslant Y$ contribute a total $O_\varepsilon(Y^3 B^{1+\varepsilon} Y^{-1/3})$ to $N_1(F; B)$. If we sum over dyadic intervals with $Y \ll B^{1/3}$ the result is a contribution $O(B^{17/9+\varepsilon})$. This is clearly satisfactory for Theorem 6. The reader may note that it is only for cubic surfaces that the exponent $52/27$ is required. In all other cases one can do better.

4.4 Theorem 8 – Large solutions

We turn now to Theorem 8. For the remainder of this lecture, all implied constants may depend on the form G. This dependence will not be mentioned explicitly. It will be convenient to make a change of variable in G so that the coefficient of x^d is positive. This will not affect the result at all.

The reader may care to note that our treatment differs somewhat from that presented in [15, §7]. This is because we have made the simplifying assumption that the binary form G is irreducible over $\mathbb{Q}$.

One technical difficulty with the proof of Theorem 8 is that one may have a value of $G(x, y)$ in the range $1 \leqslant G(x, y) \leqslant X$, for which x, y are considerably larger than $X^{1/d}$. The first task is to show that this happens relatively rarely. We assume that $C \gg X^{1/d}$ and define

$$S(X, C) =$$
$$\#\{(x, y) \in \mathbb{Z}^2 : 1 \leqslant G(x, y) \leqslant X,\ C < \max(|x|, |y|) \leqslant 2C,\ \text{h.c.f.}(x, y) = 1\}.$$

Now, if x, y is counted by $S(X, C)$ then there is at least one factor $x - ay$ of $G(x, y)$ for which $|x - ay| \ll X^{1/d}$. Hence if we take $C \geqslant cX^{1/d}$ with a sufficiently large constant c, we must have

$$C \ll |x - a'y| \ll C$$

for every other factor $x - a'y$ of $G(x, y)$. It then follows that

$$|x - ay| \ll XC^{1-d}. \tag{4.5}$$

Since Roth's Theorem implies that $|x - ay| \gg_\varepsilon C^{-1-\varepsilon}$ we deduce that $C \ll X^2$. (In fact we may draw a stronger conclusion, but the above suffices.) Thus we will have $S(X, C) = 0$ unless $C \ll X^2$.

We now estimate the contribution to $S(X, C)$ arising from pairs (x, y) for which (4.5) holds with a particular value of a. Such pairs produce primitive lattice points in the parallelogram $|y| \leqslant 2C$, $|x - ay| \ll XC^{1-d}$. In general a parallelogram of area A, centred on the origin, will contain $O(1 + A)$ primitive lattice points (see [15, Lemma 1, part (vii)]). We therefore deduce that

$$S(X, C) \ll 1 + XC^{2-d}.$$

We proceed to sum this up, for dyadic ranges with $C \ll X^2$. Thus, if

$$S'(X, C) =$$
$$\#\{(x, y) \in \mathbb{Z}^2 : 1 \leqslant G(x, y) \leqslant X,\ \max(|x|, |y|) > C,\ \text{h.c.f.}(x, y) = 1\}$$

we will deduce that
$$S'(X, C) \ll \log X + XC^{2-d},$$

when $C \gg X^{1/d}$.

We proceed to define

$$r(n) = \#\{(x, y) \in \mathbb{Z}^2 : n = G(x, y)\}$$

and

$$r_1(n; C) = \#\{(x, y) \in \mathbb{Z}^2 : n = G(x, y), \ \max(|x|, |y|) \leqslant C\},$$

$$r_2(n; C) = \#\{(x, y) \in \mathbb{Z}^2 : n = G(x, y), \ \max(|x|, |y|) > C\}.$$

Thus

$$\sum_{n \leqslant X} r_2(n; C) = \sum_{h \ll X^{1/d}} S'\left(\frac{X}{h^d}, \frac{C}{h}\right)$$

$$\ll \sum_{h \ll X^{1/d}} \left\{\log X + \frac{X}{h^d}\left(\frac{C}{h}\right)^{2-d}\right\}$$

$$\ll X^{1/d} \log X + X C^{2-d} \sum_{h \ll X^{1/d}} h^{-2}$$

$$\ll X^{1/d} \log X + X C^{2-d}. \tag{4.6}$$

This shows that 'large' pairs (x, y) make a relatively small contribution in our problem.

It is trivial that

$$\sum r_1(n; C) \ll C^2, \tag{4.7}$$

whence

$$\sum_{n \leqslant X} r(n) \ll C^2 + X^{1/d} \log X + X C^{2-d} \ll X^{2/d},$$

providing that we choose $C = cX^{1/d}$ with a suitable constant c. It follows in particular that there are $O(X^{2/d})$ positive integers $n \leqslant X$ represented by G.

4.5 Theorem 8 – Inequivalent representations

For any n which has two inequivalent representations by the form $G(x, y)$, we must either have $r_2(n; C) > 0$, or we will produce a point (x_1, x_2, x_3, x_4) on the surface

$$E(\mathbf{x}) = G(x_1, x_2) - G(x_3, x_4) = 0,$$

in the cube $|x_i| \leqslant C$, and such that (x_1, x_2) and (x_3, x_4) are not related by an automorphism. We define $\mathcal{N}(C)$ to be the number of such points. We shall show that

$$\mathcal{N}(C) \ll_\varepsilon C^{52/27+\varepsilon}, \tag{4.8}$$

and that the form G has $O(1)$ automorphisms. However before proving this, we show how Theorem 8 will follow.

We first observe that (4.7) and (4.8) imply the estimate

$$\sum_{n \leqslant X} r_1(n; C)^2 \leqslant \mathcal{N}(C) + O\left(\sum_{n \leqslant X} r_1(n; C)\right) \ll C^2,$$

where the second sum counts pairs (x_1, x_2) and (x_3, x_4) which are related by one of the finitely many automorphisms. If $C = cX^{1/d}$ with a sufficiently small constant c, then $\max(|x|, |y|) \leqslant C$ implies $|G(x, y)| \leqslant X$. Since we are assuming that $G(1, 0) > 0$, it is then trivial that

$$\sum_{n \leqslant X} r_1(n; C) \gg C^2$$

since a positive proportion of the pairs x, y with $\max(|x|, |y|) \leqslant C$ will have $G(x, y) > 0$. Now Cauchy's inequality yields

$$\sum_{\substack{n \leqslant X \\ r_1(n; C) > 0}} 1 \geqslant \frac{\left\{ \sum_{n \leqslant X} r_1(n; C) \right\}^2}{\sum_{n \leqslant X} r_1(n; C)^2} \gg C^2.$$

We therefore see that the number of positive integers $n \leqslant X$ which are represented by G, has exact order $X^{2/d}$, as required for Theorem 8.

It remains to give a non-trivial bound for the number of integers n with two or more essentially different representations by G. Since such integers are either counted by $\mathcal{N}$ or have $r_2(n; C) > 0$, we see that the number of them is

$$\leqslant \mathcal{N}(C) + \sum_{n \leqslant X} r_2(n; C)$$

$$\ll_\varepsilon C^{52/27+\varepsilon} + X^{1/d} \log X + X C^{2-d},$$

by (4.6) and (4.8). We therefore obtain a bound $O_\varepsilon\left(X^{52/(27d-2)+\varepsilon}\right)$ on taking $C = X^{27/(27d-2)}$. This proves Theorem 8, subject to the claims made above.

4.6 Theorem 8 – Points on the surface $E = 0$

It remains to investigate integral points on the surface

$$E(\mathbf{x}) = G(x_1, x_2) - G(x_3, x_4) = 0,$$

for which $\max |x_i| \leqslant C$. Using the fact that G is irreducible one may show (see [15, §7]) that E has no rational linear or quadratic factor. Thus one may apply Theorem 6 to each factor of $E(\mathbf{x})$ to deduce that

$$N_1(E; Y) \ll_\varepsilon Y^{52/27+\varepsilon}.$$

We now write $\mathcal{N}^{(*)}(C)$ to denote the number of integral zeros of E, not necessarily primitive, lying in the cube $|x_i| \leqslant C$, but not on any line in the surface $E = 0$. Then

$$\mathcal{N}^{(*)}(C) = 1 + \sum_{h \ll C} N_1(E; B/h)$$

$$\ll_\varepsilon 1 + \sum_h (C/h)^{52/27+\varepsilon}$$

$$\ll_\varepsilon C^{52/27+\varepsilon}.$$

Points which do not lie on lines in the surface $E = 0$ therefore make a contribution which is satisfactory for (4.8).

Since G has no repeated factors, the surface E is non-singular. Thus Colliot-Thélène's result, Theorem 12, shows that E contains finitely many lines L, say. Now if L is not defined over $\mathbb{Q}$ it can have at most $O(C)$ integral points in the cube $\max |x_i| \leqslant C$. Such lines therefore make a satisfactory total contribution $O(C)$ to (4.8).

For any line L lying in the surface $E(\mathbf{x}) = 0$ one can show that either all points on L satisfy $G(x_1, x_2) = G(x_3, x_4) = 0$, or that the points on L may be written as $(x, y, a_1 x + a_2 y, a_3 x + a_4 y)$ with $a_1 a_4 \neq a_2 a_3$, so that one has

$$G(a_1 x + a_2 y, a_3 x + a_4 y) = G(x, y) \tag{4.9}$$

identically. In the first case the points on L correspond to solutions of $G(x, y) = n$ with $n = 0$, which is excluded. In the second case we produce an automorphism of G, and if L is defined over $\mathbb{Q}$ this will be a rational automorphism. Thus inequivalent solutions of $G(x, y) = n > 0$ cannot lie on rational lines in the surface $E = 0$.

Finally we note that all automorphisms produce lines in the surface, as in (4.9). Thus there can be finitely many such automorphisms.

5 Fifth lecture. Affine varieties

5.1 Theorem 15 – The exponent set $\mathcal{E}$

In this lecture we shall prove Theorem 15, which concerns the counting function (2.12) for integer points on affine varieties. We shall also illustrate Theorem 15 by deriving a new result on the representation of k-free numbers by polynomials.

We begin by following the previous proof of Theorem 14. Thus if we take $B = \max(B_1, \ldots, B_n)$ then we may use an auxiliary prime p in the range $P \ll p \ll P$, providing that $P \gg \log^2(\|F\|B)$. We will take $(t_1, \ldots, t_n)$ to be a non-singular point on the variety $F(\mathbf{t}) \equiv 0 \pmod{p}$. We then aim to find a polynomial $F_j(\mathbf{x})$, determined by $\mathbf{t}$, such that $F_j(\mathbf{x}) = 0$ for all integral points $\mathbf{x} \equiv \mathbf{t} \pmod{p}$ that are counted by $N(F; \mathbf{B})$. This will be possible if P is sufficiently large, and we will then be able to take $k \ll P^{n-1} \log(\|F\|B)$ in Theorem 15, since there are $O(p^{n-1})$ possible vectors $\mathbf{t}$ modulo p. The proofs of Lemmas 2 and 3 now go through just as before.

One major difference in our new situation lies in the fact that the values of B_i may be of unequal magnitudes. Thus one cannot arrange for F to have a non-zero term in x_n^d, say, merely by using a linear change of variables, since this could radically alter the sizes of the B_i. We therefore use a different choice for the exponent set $\mathcal{E}$. The set $\mathcal{E}$ must firstly allow us still to conclude that $F(\mathbf{x})$ does not divide $F_j(\mathbf{x})$, and secondly permit us to bound the determinant Δ as sharply as possible.

To achieve the first goal, we write

$$F(X_1,\ldots,X_n) = \sum_{\mathbf{f}} a_{\mathbf{f}} X_1^{f_1} \ldots X_n^{f_n},$$

and let $P(F)$ be the Newton polyhedron of F, defined as the convex hull of the points $\mathbf{f} \in \mathbb{R}^n$ for which $a_{\mathbf{f}} \neq 0$. There may be more than one exponent vector $\mathbf{f}$ for which $\mathbf{B}^{\mathbf{f}}$ takes the maximal value T, but in any case the maximum will be achieved for at least one vertex $\mathbf{m}$, say of P, so that

$$T = \mathbf{B}^{\mathbf{m}}. \tag{5.1}$$

We shall then define our exponent set $\mathcal{E}$ to be

$$\mathcal{E} = \left\{ (e_1,\ldots,e_n) \in \mathbb{Z}^n \, : \, e_i \geqslant 0, \ (1 \leqslant i \leqslant n), \ \sum_{i=1}^{n} e_i \log B_i \leqslant Y, \right.$$
$$\left. e_i < m_i \text{ for at least one } i \right\}. \tag{5.2}$$

We now show that the choice (5.2) ensures that $F(\mathbf{x}) \nmid F_j(\mathbf{x})$. For any two polynomials $G_1, G_2 \in \mathbb{R}[\mathbf{x}]$ we have $P(G_1 G_2) = P(G_1) + P(G_2)$, (Ostrowski [27]). Now, for any polynomial G, the set $P(F) + P(G)$ must contain a point $\mathbf{m} + \mathbf{g}$ for some exponent vector $\mathbf{g}$ of G. Since $\mathbf{g}$ has non-negative components it follows that $\mathbf{m} + \mathbf{g} \notin \mathcal{E}$. Thus we cannot have $F_j = FG$, whence F cannot divide F_j, as required.

5.2 Completion of the proof of Theorem 15

It remains to estimate the size of the determinant Δ and to compute the exponent ν for our new set $\mathcal{E}$.

Tackling the determinant Δ first, we use the fact that $\left| x_i^{(j)} \right| \leqslant B_i$, to show that the column corresponding to the exponent vector $\mathbf{e}$ consists of elements of modulus at most

$$\mathbf{B}^{\mathbf{e}} = \prod_{i=1}^{n} B_i^{e_i}.$$

It follows that

$$|\Delta| \leqslant E^E \mathbf{B}^{\mathbf{E}}, \tag{5.3}$$

where

$$\sum_{\mathbf{e} \in \mathcal{E}} \mathbf{e} = \mathbf{E}. \tag{5.4}$$

For any $\mathbf{e} \in \mathbb{Z}^n$ with $e_i \geqslant 0$ we set

$$\sigma(\mathbf{e}) = \sum_{i=1}^{n} e_i \log B_i.$$

Then

$$\log \mathbf{B}^{\mathbf{E}} = \sum_{\mathbf{e} \in \mathcal{E}} \sigma(\mathbf{e}) = \sum_{\sigma(\mathbf{e}) \leqslant Y} \sigma(\mathbf{e}) - \sum{}^{(1)} \sigma(\mathbf{e}),$$

where $\sum^{(1)}$ denotes the conditions $\sigma(\mathbf{e}) \leqslant Y$ and $e_i \geqslant m_i$ for $1 \leqslant i \leqslant n$. If we substitute $e_i + m_i$ for e_i in $\sum^{(1)}$ we obtain

$$\log \mathbf{B}^{\mathbf{E}} = \sum_{\sigma(\mathbf{e}) \leqslant Y} \sigma(\mathbf{e}) - \sum_{\sigma(\mathbf{e}) \leqslant Y - \log T} (\sigma(\mathbf{e}) + \log T)$$

$$= \sum_{Y - \log T < \sigma(\mathbf{e}) \leqslant Y} \sigma(\mathbf{e}) - (\log T) \sum_{\sigma(\mathbf{e}) \leqslant Y - \log T} 1$$

$$= \{Y + O(\log T)\} \sum_{Y - \log T < \sigma(\mathbf{e}) \leqslant Y} 1 - (\log T) \sum_{\sigma(\mathbf{e}) \leqslant Y - \log T} 1. \tag{5.5}$$

Now, if $B = \max B_i$, then

$$\#\{\mathbf{e} : \sigma(\mathbf{e}) \leqslant Z\} = \frac{Z^n}{n! \prod \log B_i} + O\left(\frac{Z^{n-1}}{\prod \log B_i} \log B\right) \tag{5.6}$$

for $Z \geqslant 0$, as an easy induction on n shows. Thus

$$(\log T) \sum_{\sigma(\mathbf{e}) \leqslant Y - \log T} 1 = \frac{Y^n}{n! \prod \log B_i} \log T + O\left(\frac{Y^{n-1}}{\prod \log B_i} \log^2 T\right) \tag{5.7}$$

whether $Y \geqslant \log T$ or not.

It is not so easy to estimate the first term in (5.5). However for any $\delta \in (0, 1]$ we have

$$\int_Z^{Z(1+\delta)} \left(\sum_{Y - \log T < \sigma(\mathbf{e}) \leqslant Y} 1\right) dY = \sum_{\mathbf{e}} \int 1 \, dY,$$

with the range of integration on the right being for $Z \leqslant Y \leqslant Z(1 + \delta)$ and $\sigma(\mathbf{e}) \leqslant Y \leqslant \sigma(\mathbf{e}) + \log T$. Thus (5.6) yields

$$\int_Z^{Z(1+\delta)} \left(\sum_{Y - \log T < \sigma(\mathbf{e}) \leqslant Y} 1\right) dY \leqslant (\log T) \sum_{Z - \log T \leqslant \sigma(\mathbf{e}) \leqslant Z(1+\delta)} 1$$

$$= \frac{\log T}{\prod \log B_i} \left\{\frac{(1+\delta)^n - 1}{n!} Z^n + O(Z^{n-1} \log T)\right\}.$$

Similarly we find that

$$\int_Z^{Z(1+\delta)} \left(\sum_{Y-\log T < \sigma(\mathbf{e}) \leqslant Y} 1 \right) dY \geqslant (\log T) \sum_{Z \leqslant \sigma(\mathbf{e}) \leqslant Z(1+\delta) - \log T} 1$$

$$= \frac{\log T}{\prod \log B_i} \left\{ \frac{(1+\delta)^n - 1}{n!} Z^n + O(Z^{n-1} \log T) \right\}.$$

Thus there is some Y in the range $Z \leqslant Y \leqslant Z(1+\delta)$ for which

$$\sum_{Y-\log T < \sigma(\mathbf{e}) \leqslant Y} 1 = \frac{\log T}{\prod \log B_i} \left\{ \frac{(1+\delta)^n - 1}{\delta n!} Z^{n-1} + O(\delta^{-1} Z^{n-2} \log T) \right\}$$

$$= \frac{\log T}{\prod \log B_i} \left\{ \frac{Z^{n-1}}{(n-1)!} + O(\delta Z^{n-1}) + O(\delta^{-1} Z^{n-2} \log T) \right\}$$

$$= \frac{\log T}{\prod \log B_i} \left\{ \frac{Y^{n-1}}{(n-1)!} + O(\delta Y^{n-1}) + O(\delta^{-1} Y^{n-2} \log T) \right\}.$$

If $Y \geqslant \log T$ we may choose

$$\delta = \sqrt{\frac{\log T}{Y}}$$

so that

$$\sum_{Y-\log T < \sigma(\mathbf{e}) \leqslant Y} 1 = \frac{\log T}{\prod \log B_i} \left\{ \frac{Y^{n-1}}{(n-1)!} + O(Y^{n-3/2} (\log T)^{1/2}) \right\}. \quad (5.8)$$

If $Z \geqslant \log T$, we now see that any range $Z \leqslant Y \leqslant 2Z$ contains a value of Y for which (5.8) holds. For such Y equations (5.5), (5.7) and (5.8) yield

$$\log \mathbf{B^E} = \frac{\log T}{\prod \log B_i} \frac{Y^n}{(n-1)!} - \frac{\log T}{\prod \log B_i} \frac{Y^n}{n!} + O\left(\frac{\log T}{\prod \log B_i} Y^{n-1/2} (\log T)^{1/2} \right)$$

$$= \frac{n-1}{n!} \frac{\log T}{\prod \log B_i} Y^n \left\{ 1 + O\left(\sqrt{\frac{\log T}{Y}} \right) \right\}. \quad (5.9)$$

A similar but simpler argument provides us with our estimate for ν. Indeed (3.8) and (3.9) still apply, but with n replaced by $n+1$. It follows that

$$\nu = \frac{1}{n(n-2)!} \{(n-1)! E\}^{n/(n-1)} \{1 + O(E^{-1/(n-1)})\},$$

where $E = \#\mathcal{E}$. However, as in (5.5), we have

$$\#\mathcal{E} = \sum_{\sigma(\mathbf{e}) \leqslant Y} 1 - \sum_{\sigma(\mathbf{e}) \leqslant Y - \log T} 1 = \sum_{Y - \log T < \sigma(\mathbf{e}) \leqslant Y} 1.$$

For the value of Y we have chosen, it follows from (5.8) that

$$E = \frac{\log T}{\prod \log B_i} \frac{Y^{n-1}}{(n-1)!} \left\{ 1 + O\left(\sqrt{\frac{\log T}{Y}} \right) \right\}. \tag{5.10}$$

Since $\log B_i \leqslant \log T$ for each index i, we deduce that

$$E \gg \left(\frac{Y}{\log T} \right)^{n-1}$$

if $Y \gg \log T$, and hence that

$$\nu = \frac{Y^n}{n(n-2)!} \left(\frac{\log T}{\prod \log B_i} \right)^{n/(n-1)} \left\{ 1 + O\left(\sqrt{\frac{\log T}{Y}} \right) \right\}. \tag{5.11}$$

As before, we need $p^\nu > |\Delta|$. A comparison of (5.3), (5.9) and (5.11) now shows that if $E \leqslant C$, say, then we can take

$$\log p = \frac{\log \mathbf{B^E}}{\nu} + O_C(1) = \left(\frac{\prod \log B_i}{\log T} \right)^{1/(n-1)} \left\{ 1 + O\left(\sqrt{\frac{\log T}{Y}} \right) \right\} + O_C(1).$$

If $\varepsilon > 0$ is such that $B_i \geqslant B^\varepsilon$ for $1 \leqslant i \leqslant n$, and $Y = \lambda \log T$, then (5.10) yields $E \ll_{\lambda, \varepsilon, d} 1$. By allowing λ to become arbitrarily large we can therefore take

$$\log p \leqslant (1 + \varepsilon) \left(\frac{\prod \log B_i}{\log T} \right)^{1/(n-1)}.$$

Thus the bound given in Theorem 15 certainly holds if $B_i \geqslant B^\varepsilon$ for $1 \leqslant i \leqslant n$.

On the other hand, if $B_1 \leqslant B^\varepsilon$, say, we can merely take the polynomials $F_j(\mathbf{x})$ as $F_j(\mathbf{x}) = x_1 - a_j$ for the various integers $a_j \in [-B_1, B_1]$. Clearly each relevant point will be a zero of such a polynomial, and there are $\ll B_1 \ll T^\varepsilon$ of them. This completes the proof of Theorem 15.

5.3 Power-free values of polynomials

We shall now apply Theorem 15 to the problem of power-free values of polynomials. Let $f[X] \in \mathbb{Z}[X]$ be an irreducible polynomial of degree d. It can happen that there is a prime p such that $p^2 \mid f(n)$ for every integer n, even though f is necessarily primitive. (The polynomial $X^4 + 2X^3 - X^2 - 2X + 4$ provides an example, with $p = 2$.) However if we assume that there is no prime p such that $p^2 \mid f(n)$ for every integer n, then we would expect $f(n)$ to represent infinitely many square-free integers. This is known only for polynomials of degree $d \leqslant 3$. This is relatively trivial for $d \leqslant 2$, and was shown by Erdős [6] for $d = 3$. When one considers polynomials of higher degree one may ask for which values of k one can assert that $f(n)$ is infinitely often k-th power free, or "k-free" for short. Hooley showed that one may take $k = d - 1$ in general, and Nair [25] that any $k \geqslant (\sqrt{2} - \frac{1}{2})d$ is admissible.

We shall strengthen this result for $d \geqslant 10$ as follows.

Theorem 16 *Let $f[X] \in \mathbb{Z}[X]$ be an irreducible polynomial of degree d, with positive leading coefficient. Suppose we have an integer $k \geqslant (3d + 2)/4$, and assume moreover that there is no prime p such that $p^k \mid f(n)$ for every integer n. Then*

$$\#\{n \leqslant B : f(n) \text{ is } k\text{-free}\} \sim C(k, f)B$$

as B tends to infinity, where

$$C(k, f) = \prod_p \left(1 - \rho(p^k)/p^k\right),$$

with $\rho(n)$ being the number of zeros modulo n of the polynomial $f(x)$.

Note that although $(3d+2)/4 < \left(\sqrt{2} - \frac{1}{2}\right)d$ as soon as $d \geqslant 4$, we only have $\lceil (3d+2)/4 \rceil < \lceil \left(\sqrt{2} - \frac{1}{2}\right) \rceil d$ when $d \geqslant 10$.

The initial stages of the argument are straightforward. We shall assume that f is fixed, so that any order constants in the following may depend on f.

One has

$$\sum_{h^k \mid f(n)} \mu(h) = \begin{cases} 1, & f(n) \text{ is } k\text{-free}, \\ 0, & \text{otherwise}, \end{cases}$$

whence

$$\#\{n \leqslant B : f(n) \text{ is } k\text{-free}\} = \sum_h \mu(h) N(h, B),$$

with

$$N(h, B) = \#\{n \leqslant B : h^k \mid f(n)\}.$$

Moreover we see that $N(h, B) = 0$ for $h \gg B^{d/k}$, and that

$$N(h, B) = \frac{B}{h^k} \rho(h^k) + O\left(\rho(h^k)\right)$$

in general. Since $\rho(h)$ is multiplicative, and $\rho(p) \ll 1$, we see that $\rho(h^k) \ll_\varepsilon h^\varepsilon$ for any $\varepsilon > 0$. Here, and for the rest of this section, we restrict h to square-free values. It follows that

$$\sum_{h \leqslant H} \mu(h) N(h, B) = B \sum_{h \leqslant H} \mu(h)\rho(h)/h^k + O_\varepsilon\left(\sum_{h \leqslant H} h^\varepsilon\right)$$

$$= B\{C(k, f) + O_{f,\varepsilon}(H^{\varepsilon-k})\} + O_{f,\varepsilon}(H^{1+\varepsilon}).$$

Thus, if we choose $H = B^{1-\delta}$ with any fixed positive δ we will have

$$\sum_{h \leqslant B^{1-\delta}} \mu(h) N(h, B) \sim C(k, f)B.$$

It therefore remains to show that

$$\sum_{B^{1-\delta} < h \ll B^{d/k}} N(h, B) = o(B). \tag{5.12}$$

For any integer H with $B^{1-\delta} \ll H \ll B^{d/k}$, we have

$$\sum_{H < h \leqslant 2H} N(h, B) \leqslant N(F; B_1, B_2, B_3),$$

where $F(x_1, x_2, x_3) = f(x_1) - x_2^k x_3$ and $B_1 = B$, $B_2 = 2H$, $B_3 \ll B^d/H^k$. This polynomial F is clearly absolutely irreducible. It will be convenient to replace the above inequality by a slightly sharper one

$$\sum_{H < h \leqslant 2H} N(h, B) \leqslant N'(F; B_1, B_2, B_3), \tag{5.13}$$

in which $N'(F; B_1, B_2, B_3)$ counts only solutions in which x_2 is a positive square-free integer. We now apply Theorem 15, in which T will be of order B^d. Writing

$$\eta = \frac{\log H}{\log B}$$

so that

$$1 - \delta \leqslant \eta \leqslant d/k + o(1)$$

we find that all relevant triples satisfy one of

$$O_{d,\varepsilon}\left(B^{2\sqrt{\eta(1-k\eta/d)}+\varepsilon}\right) \tag{5.14}$$

auxiliary equations $F_j(x_1, x_2, x_3) = 0$, of degree at most $D = D(d, \varepsilon)$. On multiplying by x_2^{kD} we can eliminate x_3 from such an auxiliary equation, using the fact that

$$x_2^k x_3 = f(x_1). \tag{5.15}$$

It follows that we may assume that the auxiliary equations take the shape $G_j(x_1, x_2) = 0$, with no dependence on x_3. This elimination process will produce polynomials G_j which do not vanish identically, since the original polynomials F_j were coprime to F. Moreover the degree of each G_j is at most $D(d, \varepsilon)$ for some new function D. If G_j is not absolutely irreducible we may split it into its irreducible factors. In view of Theorem 3 we can concentrate on factors which have rational coefficients. Thus we may suppose that $G_j(x_1, x_2)$ is absolutely irreducible.

We now write $H_j(y_1, y_2, y_3) = y_3^D G_j(y_1/y_3, y_2/y_3)$, so that H_j is a non-zero form of degree D. According to Theorem 5 there are two cases. In the first case $N(H_j; L) \leqslant D^2$ where $L = \max(B, 2H, 1)$. It then follows that j determines $O_{d,\varepsilon}(1)$ admissible values for x_1, x_2. Moreover, since we are assuming that $x_2 > 0$, the relation (5.15) then shows that any admissible pair x_1, x_2 determines at most one admissible value of x_3. In the second case the coefficients of G_j may be taken to be integers of size $O(B^N)$ for some exponent N

depending at most on d and ε. Since we are only interested in values $x_2 > 0$ we may remove any factors of x_2 from the polynomial G_j. Thus we can write

$$G_j(X_1, X_2) = G_j^{(0)}(X_1) + X_2\, G_j^{(1)}(X_1, X_2)$$

for appropriate integral polynomials $G_j^{(0)}$ and $G_j^{(1)}$, with $G_j^{(0)}$ not identically zero, and having coefficients $O(B^N)$. We now subdivide this second case into two subcases. In the first subcase we will have $f(x) \nmid G_j^{(0)}(x)$, while in the second subcase we will have $f(x) \mid G_j^{(0)}(x)$.

For the first of these subcases we observe that x_2 must divide $G_j^{(0)}(x_1)$. However it is clear from (5.15) that x_2 also divides $f(x_1)$. Thus we conclude that x_2 must be a factor of the resolvent R of f and $G_j^{(0)}$. Since $f(X)$ does not divide $G_j^{(0)}(X)$ it follows that R is non-zero, and our bound on the coefficients of $G_j^{(0)}$ implies that $R \ll B^{N'}$ for some N' depending only on d and ε. Hence x_2 can take at most $d(R) \ll_\varepsilon B^\varepsilon$ values in this first subcase. (Here $d(\ldots)$ is the usual divisor function.) Now x_2 is assumed to be square-free; so that $\rho(x_2) \ll B^\varepsilon$. Moreover the equation (5.15) implies that $f(x_1) \equiv 0 \pmod{x_2}$. Hence we see that the number of possible values of x_1 corresponding to each admissible x_2 is

$$\ll (1 + B/x_2)\rho(x_2) \ll_\varepsilon (1 + B/H)B^\varepsilon \ll_\varepsilon B^{\delta+\varepsilon}.$$

As before x_1 and x_2 determine x_3, by (5.15). It therefore follows that each value of j produces $O(B^{\delta+2\varepsilon})$ triples (x_1, x_2, x_3) in the first subcase.

Turning to the second subcase, in which $F(X) \mid G_j^{(0)}(X)$, we note that $G_j^{(0)}$ must have degree at least d. We then apply Theorem 15 to the equation $G_j(x_1, x_2) = 0$, taking $B_1 = B$, $B_2 = 2H$, and $T \geqslant B^d$. This shows that there are $O_\varepsilon(B^\varepsilon H^{1/d})$ possible pairs x_1, x_2, and as usual any such pair determines at most one value of x_3.

We can finally conclude, via (5.14), that (5.13) holds with

$$N'(F; B_1, B_2, B_3) \ll B^{2\sqrt{\eta(1-k\eta/d)}+\eta/d+2\varepsilon},$$

providing that δ is taken to be small enough. This suffices to establish (5.12), assuming that we have

$$\sup_{1 \leqslant \eta \leqslant d/k} 2\sqrt{\eta(1 - k\eta/d)} + \eta/d < 1.$$

Since this holds providing that $k \geqslant (3d + 2)/4$, the proof of Theorem 16 is complete.

6 Sixth lecture. Sums of powers, and parameterizations

In this lecture we shall look at the projective surface

$$x_1^d + x_2^d = x_3^d + x_4^d \tag{6.1}$$

and the affine surface

$$x_1^d + x_2^d + x_3^d = N.$$

It transpires that our basic techniques are well adapted to these, but need to be supplemented by information about possible curves of low degree (or low genus) lying on these varieties.

6.1 Theorem 13 – Equal sums of two powers

We shall begin by examining the surface (6.1). We can make a direct application of Theorem 14 to show that all points in the box $\max |x_i| \leqslant B$ must lie on one of $O_{\varepsilon,d}(B^{3\sqrt{d}+\varepsilon})$ curves in the surface. Moreover Theorem 12 tells us that $O_d(1)$ of these curves can have degree $d-2$ or lower.

We begin by disposing of points which lie on a curve C of degree $k \geqslant d-1$ in the surface (6.1). Here we can apply Theorem 4 to show that any such curve contributes $O_{\varepsilon,k}(B^{2/k+\varepsilon})$ to $N_1(F;B)$. Since we have $d-1 \leqslant k \ll_{d,\varepsilon} 1$ it follows that the total contribution to $N_1(F;B)$ from all such curves arising from Theorem 14 will be $O_\varepsilon(B^{3/\sqrt{d}+2/(d-1)+\varepsilon})$, as required for Theorem 13.

We therefore turn our attention to the curves of degree at most $d-2$. Theorem 12 assures us that we can produce a finite list of these, independently of B. Thus we do not have the usual problem of uniformity with respect to B. It follows that we may apply the theorems (1.3) and (1.4) of Néron and Faltings to show that any curve of genus 1 will contribute $O_{\varepsilon,d}(B^\varepsilon)$ and any curve of genus 2 or more will contribute $O_d(1)$. Thus we have a total contribution of $O_{\varepsilon,d}(B^\varepsilon)$ to $N_1(F;B)$.

It follows that we must now examine the possibility that there are curves of genus zero on the surface (6.1). In doing this it in fact suffices to work over $\mathbb{C}$. Since any curve of genus zero can then be parameterized by polynomials, we shall look for possible polynomial solutions to (6.1). This is clearly a question of interest in its own right, in view of the solutions (1.6) and (1.8) for the cases $d=3$ and $d=4$.

We shall establish the following general result, following an argument due to Newman and Slater [26].

Lemma 5 *Let $n \geqslant 2$ and let $f_1(t), \ldots, f_n(t) \in \mathbb{C}[t]$ be non-zero polynomials. Suppose that $d \geqslant n(n-2)$. Then if*

$$\sum_{j=1}^n f_j(t)^d = 0$$

holds identically, there must be two polynomials f_i, f_j which are proportional to each other.

If one of the polynomials is constant it suffices to have $d \geqslant (n-2)(n-1)$.

We therefore see that there can be no analogue of Euler's parametric solution to (1.7) for the surfaces (6.1) when the degree is 8 or more. Indeed, when $n = 4$ and $d \geqslant 8$ we may conclude that $f_4(t) = cf_3(t)$, say. Then either $c^d = -1$, or

$$f_1(t)^d + f_2(t)^d + \widetilde{f}_3(t)^d = 0$$

with

$$\widetilde{f}_3(t) = \left(1 + c^d\right)^{1/d} f_3(t) \neq 0.$$

In the first case we must have $f_1(t)^d + f_2(t)^d = f_3(t)^d + f_4(t)^d = 0$. In the second case, Lemma 5 shows that at least two of $f_1(t)$, $f_2(t)$, $\widetilde{f}_3(t)$ are proportional, and hence that all three are. In either case we see that the original polynomials $f_1(t), \ldots, f_4(t)$ are all proportional. These polynomials would then not parameterize a curve. We therefore see that, for $d \geqslant 8$, any curve of genus zero lying in the surface

$$x_1^d + x_2^d - x_3^d - x_4^d = 0$$

must be one of the obvious lines, and hence cannot contribute to $N_1(F; B)$. This establishes Theorem 13 when $d \geqslant 8$. On the other hand, if $d \leqslant 7$ one has $3/\sqrt{d} + 2/(d-1) \geqslant 1$, so that Theorem 13 follows from Theorem 11 in this case.

We now prove Lemma 5, which will be done by induction on n, the result being trivial for $n = 2$. Clearly we can suppose that the polynomials $f_j(t)$ have no common factor. It will be convenient to write $F_j(t) = f_j(t)^d$. We begin by differentiating the relation

$$\sum_{j=1}^{n} F_j(t) = 0$$

repeatedly, and we set

$$H_{ij}(t) = F_j^{-1}(t)\left(\frac{\mathrm{d}}{\mathrm{d}t}\right)^i F_j(t) \quad (0 \leqslant i \leqslant n - 2). \tag{6.2}$$

We then deduce a system of equations

$$\sum_{j=1}^{n} H_{ij}(t) F_j(t) = 0 \quad (0 \leqslant i \leqslant n - 2),$$

which we write in matrix form as $H\mathbf{F} = \mathbf{0}$, where H is the $(n-1) \times n$ matrix with entries $H_{ij}(t)$, and $\mathbf{F}$ is the column vector of length n, with entries $F_j(t)$.

We consider two cases. Suppose firstly that H has rank strictly less than $n - 1$. In this case all the $(n-1) \times (n-1)$ minors $H_j(t)$ (say) must vanish. We now observe that

$$F_1(t)F_2(t)\ldots F_{n-1}(t)H_n(t) = \begin{vmatrix} F_1(t) & \cdots & F_{n-1}(t) \\ \cdots & \cdots & \cdots \\ \left(\frac{\mathrm{d}}{\mathrm{d}t}\right)^{n-2}F_1(t) & \cdots & \left(\frac{\mathrm{d}}{\mathrm{d}t}\right)^{n-2}F_{n-1}(t) \end{vmatrix},$$

which is the Wronskian of $F_1(t), \ldots, F_{n-1}(t)$. According to our assumption this vanishes, and hence the polynomials $F_1(t), \ldots, F_{n-1}(t)$ will be linearly dependent over $\mathbb{C}$. There is therefore a set $\mathcal{S} \subseteq \{1, \ldots, n-1\}$ for which we have a relation

$$\sum_{j \in \mathcal{S}} \alpha_j F_j(t) = 0$$

in which none of the α_j are zero. Moreover, if $\#\mathcal{S} = n'$, we will have $2 \leqslant n' \leqslant n-1$. We can now pick any d-th roots $\alpha^{1/d}$ to obtain an equation of the form

$$\sum_{j \in \mathcal{S}} \left\{ \alpha_j^{1/d} f_j(t) \right\}^d = 0.$$

According to our induction hypothesis, two of the polynomials f_i, f_j must be proportional, as required.

We turn now to the second case, in which the rank of H is $n-1$. In this case $\mathbf{F}$ must be proportional to $(H_1, \ldots, H_n)$. Without loss of generality we can assume that the degree h, say, of $f_1(t)$ is maximal. On recalling that $F_j(t) = f_j(t)^d$ we see from (6.2) that there are polynomials $g_{ij}(t)$ such that

$$H_{ij}(t) = \frac{g_{ij}(t)}{f_j(t)^i}, \quad \deg\left(g_{ij}(t)\right) \leqslant i(h-1).$$

Consequently, if we define

$$Q(t) = \left\{ \prod_{j=1}^{n} f_j(t) \right\}^{n-2},$$

and set

$$Q(t) H_j(t) = P_j(t), \quad (1 \leqslant j \leqslant n), \tag{6.3}$$

it follows that $P_j(t)$ must be a polynomial, and that

$$\deg\left(P_j(t)\right) \leqslant (n-2)nh - \frac{(n-1)(n-2)}{2}. \tag{6.4}$$

However the polynomials $F_1(t), \ldots, F_n(t)$ will be coprime, and $\mathbf{F}(t)$ is proportional to $(P_1(t), \ldots, P_n(t))$. It follows that

$$hd = \deg\left(f_1(t)^d\right) = \deg\left(F_1(t)\right) \leqslant \deg\left(P_1(t)\right) \leqslant (n-2)nh - \frac{(n-1)(n-2)}{2}.$$

We therefore obtain a contradiction if

$$d > (n-2)\left(n - \frac{n-1}{2h}\right).$$

In particular we cannot have $d \geqslant n(n-2)$, irrespective of the value of h. If one of the polynomials is constant then the bound (6.4) may be replaced by

$$\deg\left(P_j(t)\right) \leqslant (n-2)(n-1)h - \frac{(n-1)(n-2)}{2},$$

and we obtain a contradiction when $d \geqslant (n-2)(n-1)$. This completes the proof of Lemma 5.

6.2 Parameterization by elliptic functions

Lemma 5 gives good control over possible genus zero curves on diagonal hyper-surfaces

$$X_1^d + \ldots + X_n^d = 0. \tag{6.5}$$

One can prove analogous results for genus 1 curves in general, but here we shall restrict attention to plane cubic curves. These can be parameterized using the Weierstrass elliptic function. Specifically, if there is a plane cubic curve contained in the variety (6.5), then there are functions

$$f_j(z) = A_j \wp'(z) + B_j \wp(z) + C_j, \quad (1 \leqslant j \leqslant n), \tag{6.6}$$

not all proportional to each other, such that

$$\sum_{j=1}^n f_j(z)^d = 0$$

identically for $z \in \mathbb{C}$. Since the vector $(f_1(z), \ldots, f_n(z))$ must describe a cubic curve rather than a straight line, we conclude that the matrix

$$M = \begin{pmatrix} A_1 & \ldots & A_n \\ B_1 & \ldots & B_n \\ C_1 & \ldots & C_n \end{pmatrix}$$

must have rank 3.

We now have the following result, analogous to Lemma 5.

Lemma 6 *Let $n \geqslant 2$ and let $f_1(z), \ldots, f_n(z)$ satisfy (6.6). Assume further that the corresponding matrix M has rank 3, and that $d > (7n-1)(n-2)/6$. Then if*

$$\sum_{j=1}^n f_j(z)^d = 0$$

holds identically, there must be two functions f_i, f_j which are proportional to each other.

Before proving this we note that Green [9] gives a result which is both stronger and more general than Lemma 6. He proves that it suffices to have $d \geqslant (n-1)^2$, even when the f_j are arbitrary meromorphic functions which do not all vanish simultaneously. (We will also encounter such a condition, but it causes no problem for functions of the form (6.6).) Green's argument uses Nevanlinna theory, while our proof is more explicit and self-contained.

The result given in Lemma 6 is trivial if any function $f_j(z)$ vanishes identically, so we shall assume that each such function is non-zero. At least one matrix

$$\begin{pmatrix} A_i & A_j & A_k \\ B_i & B_j & B_k \\ C_i & C_j & C_k \end{pmatrix},$$

where $i < j < k$, will be non-singular, since M has rank 3. It follows that the simultaneous equations

$$f_i(z_0) = A_i \wp'(z_0) + B_i \wp(z_0) + C_i = 0$$
$$f_j(z_0) = A_j \wp'(z_0) + B_j \wp(z_0) + C_j = 0$$
$$f_k(z_0) = A_k \wp'(z_0) + B_k \wp(z_0) + C_k = 0$$

have no solution. Hence there can be no $z_0 \in \mathbb{C}$ at which $f_i(z)$, $f_j(z)$, $f_k(z)$ all vanish.

The argument now follows that given for Lemma 5. If the matrix H has rank $n - 2$ or less, there will be two functions which are proportional. On the other hand, if the rank of H is $n - 1$, then the vector $\mathbf{F}$ will be proportional to $(H_1, \ldots, H_n)$. In particular we see that none of the H_j can vanish identically, so that we may write

$$\frac{F_i(z)}{F_j(z)} = \left(\frac{f_i(z)}{f_j(z)} \right)^d = \frac{H_i(z)}{H_j(z)} = \frac{P_i(z)}{P_j(z)}, \tag{6.7}$$

with functions $P_i(z)$, $P_j(z)$ constructed as in (6.3). This construction shows that these functions are polynomials in $\wp$ and its derivatives, and that they have poles of order at most

$$3n(n-2) + \frac{(n-1)(n-2)}{2} = \frac{(7n-1)(n-2)}{2} \tag{6.8}$$

at the origin.

In completing the proof we shall use the fact that a doubly periodic meromorphic function has the same (finite) number of poles as zeros, counted according to multiplicity, in its fundamental parallelogram. There must be some index i for which $A_i \neq 0$, since M has rank 3. For this index, $f_i(z)$ has a single pole, of order 3. Thus f_i has zeros of total multiplicity 3. Suppose that z_0 is such a zero, and has multiplicity μ say. We have already noted that the functions $f_1, \ldots, f_n$ cannot all vanish at z_0. There is therefore an index j with $f_j(z_0) \neq 0$. Since P_j is a polynomial in $\wp$ and its derivatives, it has poles only at the origin, within the fundamental parallelogram. We then see from (6.7) that P_i will have a zero of order at least μd at the point z_0. We may apply this reasoning to each of the zeros of f_i and show that P_i has zeros of total multiplicity at least $3d$ in the fundamental parallelogram. On the other hand, (6.8) provides an upper bound for the multiplicity of the poles of P_i. A comparison of these bounds yields

$$3d \leqslant 3n(n-2) + \frac{(n-1)(n-2)}{2},$$

contradicting the assumption of the lemma. This suffices for the proof.

6.3 Sums of three powers

We turn now to the affine surface

$$x_1^d + x_2^d + x_3^d = N. \tag{6.9}$$

Bearing in mind the arithmetical significance of this we will only look at solutions with $x_i > 0$. Let $r(N)$ be the number of such solutions. When $d \geqslant 2$ we easily have $r(N) \ll_{d,\varepsilon} N^{1/d+\varepsilon}$, but no improvement in the exponent $1/d$ has hitherto been given, for any value of d. The exponent is certainly best possible for $d = 2$, and for $d = 3$ it was shown by Mahler [24] that $r(N) = \Omega(N^{1/12})$. This follows by taking $N = n^{12}$ in the identity

$$(9x^4)^3 + (3xn^3 - 9x^4)^3 + (n^4 - 9x^3n)^3 = n^{12}. \tag{6.10}$$

In general such an identity must arise from an expression of a non-zero constant as a sum of three d-th powers of polynomials. A comparison of the leading terms in such an identity shows that d must be odd, while Lemma 5 shows that we must have $d < 6$. Thus there can be no analogue of Mahler's identity for higher powers, except possibly for $d = 5$. Indeed it may be conjectured that $r(N) \ll_{d,\varepsilon} N^\varepsilon$ as soon as $d \geqslant 4$. The following result goes some way towards this.

Theorem 17 *For $d \geqslant 8$ we have*

$$r(N) \ll_\varepsilon N^{\theta/d+\varepsilon}$$

where

$$\theta = \frac{2}{\sqrt{d}} + \frac{2}{d-1}.$$

Observe that we have a non-trivial bound $\theta < 1$ for $d \geqslant 8$. Note also that the theorem remains true for $d < 8$, by the trivial bound $r(N) \ll_{d,\varepsilon} N^{1/d+\varepsilon}$, since $\theta > 1$ for $d < 8$.

For the proof we begin by applying Theorem 15 to the polynomial

$$F(x_1, x_2, x_3) = x_1^d + x_2^d + x_3^d - N,$$

taking $B_1 = B_2 = B_3 = N^{1/d} = B$, say. We conclude that all relevant points lie on one of $O_\varepsilon(B^{2/\sqrt{d}+\varepsilon})$ curves, each having degree $O_\varepsilon(1)$. When such a curve has degree $D \geqslant d - 1$, it will have $O_\varepsilon(B^{2/D+\varepsilon})$ corresponding points, by Theorem 4. The total number of solutions of (6.9) in such cases is thus $O_\varepsilon(B^{\theta+\varepsilon})$, which is satisfactory. Indeed for curves in $\mathbb{A}^3$ Pila [28, Theorem A] shows that one may replace the exponent $2/d$ in our Theorem 4 by $1/d$. Thus in our situation each curve of degree $D \geqslant d - 1$ will contribute $O_\varepsilon(B^{1/D+\varepsilon})$. It follows that Pila's result allows us to improve θ to

$$\theta = \frac{2}{\sqrt{d}} + \frac{1}{d-1}.$$

We have only stated the slightly weaker result in our theorem in order to be self-contained.

Now let C be a curve of degree at most $d - 2$, contained in the surface (6.9). If $\theta : \mathbb{A}^3 \longrightarrow \mathbb{A}^3$ is the map

$$\theta(x_1, x_2, x_3) = N^{-1/d}(x_1, x_2, x_3),$$

then $\theta(C)$ is a curve of degree at most $d - 2$, lying in the non-singular surface S, given by $y_1^d + y_2^d + y_3^d = 1$. Theorem 12 has a natural affine version, which shows that there are $O(1)$ such curves, $C_1, \ldots, C_t$, say. Obviously t and the curves C_i depend only on d and not on N. Suppose that $\theta(C) = C_i$, say. Let $\pi : \mathbb{A}^3 - \{0\} \longrightarrow \mathbb{P}^2$ be the map given by $\pi(x_1, x_2, x_3) = [(x_1, x_2, x_3)]$, where $[(x_1, x_2, x_3)]$ is the point in $\mathbb{P}^2$ represented by (x_1, x_2, x_3). Since $(0, 0, 0)$ does not satisfy (6.9) it cannot lie on the curve C, and hence π gives a regular map from C into $\mathbb{P}^2$. Moreover, since $\pi(\mathbf{x}) = \pi(\theta(\mathbf{x}))$ we find that $\pi(C) = \pi(\theta(C)) = \pi(C_i)$. Thus the Zariski closure $\overline{\pi(C)}$ must be one of a finite number of curves $\overline{\pi(C_i)}$, independent of N. Write $\Gamma_i = \overline{\pi(C_i)}$, for convenience.

This is the key point in our argument. The curves C are likely to be different for different values of N, and as N varies we will encounter infinitely many different curves C. Thus it appears that we will have a problem of uniformity in N. However these curves are all 'twists', by $N^{1/d}$, of a finite number of curves C_i, and by mapping into $\mathbb{P}^2$ each of these twists gets sent to the same curve Γ_i. The uniformity issue then disappears. Of course we are left with the problem that each point in $\mathbb{P}^2$ corresponds to many different points in $\mathbb{A}^3$. However, these are scalar multiples of one another, and at most one can be a solution of (6.9) for a particular value of N.

We now begin by disposing of the case in which Γ_i is not defined over $\mathbb{Q}$. In this case the rational point $\pi(\mathbf{x})$ lies on the intersection of Γ_i and any one of its conjugates. Such an intersection has $O(1)$ points by Bézout's Theorem. Thus Γ_i contains $O(1)$ rational points. As noted above, each such point in $\mathbb{P}^2$ can correspond to at most one solution of (6.9). Thus (6.9) has $O(1)$ solutions $\mathbf{x}$ for which $\pi(\mathbf{x})$ lies on a curve Γ_i not defined over $\mathbb{Q}$. Next, if Γ_i has genus 2 or more, it will have $O(1)$ points by Faltings' Theorem (1.4), and again there are $O(1)$ corresponding solutions of (6.9). In the case in which Γ_i has genus 1, Néron's result (1.4) similarly yields $O(B^\varepsilon)$ solutions of (6.9). Thus it remains to consider the case in which Γ_i is defined over $\mathbb{Q}$ and has genus zero.

In this final case we observe that a curve of genus zero defined over $\mathbb{Q}$ can be parameterized by rational functions. To be more precise, there are forms $f_1(u, v)$, $f_2(u, v)$, $f_3(u, v) \in \mathbb{Z}[u, v]$, with no common factor, such that every rational point on Γ_i, with at most finitely many exceptions, is a non-zero rational multiple of $(f_1(u, v), f_2(u, v), f_3(u, v))$ for appropriate coprime $u, v \in \mathbb{Z}$. Thus it remains to examine solutions to the equation

$$\lambda^d \{ f_1(u, v)^d + f_2(u, v)^d + f_3(u, v)^d \} = N, \quad \lambda \in \mathbb{Q}, \ u, v \in \mathbb{Z}, \ (u, v) = 1.$$

The forms f_1, f_2, f_3 can be considered fixed, independently of N, but λ, and of course u and v, may vary. We shall need to control λ. Write $\lambda = \mu/\nu$ with

$(\mu, \nu) = 1$. We now use the fact that the forms f_i are coprime to produce relations of the form

$$\sum_{i=1}^{3} g_i(u,v) f_i(u,v) = Gu^r, \qquad \sum_{i=1}^{3} h_i(u,v) f_i(u,v) = Hv^r,$$

where $g_i(u,v)$, $h_i(u,v)$ are integral forms, and G, H are non-zero integer constants. Since $\lambda f_i(u,v)$ must be integral for $i = 1, 2, 3$, in any solution of interest, we conclude that $\nu \,|\, f_i(u,v)$, and hence that $\nu \,|\, Gu^r$ and $\nu \,|\, Hv^r$. However u and v are assumed to be coprime, so that $\nu \,|\, GH$. It follows that ν takes finitely many values. Moreover we have $\mu^d \,|\, N$, so that μ can take at most $O(N^\varepsilon)$ values.

We are left to consider the number of solutions of the Thue equation

$$f_1(u,v)^d + f_2(u,v)^d + f_3(u,v)^d = \nu^d \mu^{-d} N \tag{6.11}$$

for fixed forms f_1, f_2, f_3 and fixed μ, ν. We denote the form on the left by $F(u,v)$. If $F(u,v)$ has at least two distinct rational factors, say $F_1(u,v)$ and $F_2(u,v)$, then we will have $F_1(u,v) = N_1$ and $F_2(u,v) = N_2$ for certain factors N_1, N_2 of $\nu^d \mu^{-d} N$. These two equations determine $O(1)$ values of u, v, by elimination, so that (6.11) has $O_\varepsilon(N^\varepsilon)$ solutions. Now suppose to the contrary that F is a constant multiple of a power of an irreducible form F_1 say, in which case we have to consider solutions of an equation $F_1(u,v) = N_1$. When F_1 has degree 3 or more one may apply an old result of Lewis and Mahler [23], which shows that there are $O\big(A^{\omega(N_1)}\big)$ such solutions, with a constant A depending only on F_1. This is enough to show that there are $O_\varepsilon(N^\varepsilon)$ solutions in this case.

Now consider the case in which F_1 has degree 2. Here there will be $O_\varepsilon(N^\varepsilon)$ solutions to the equation $F_1(u,v) = N_1$ providing that we have $u, v \ll N$. However we assumed that the forms f_i had no common factor, and we may therefore take f_1, say, to be coprime to F_1. Now if $F_1(u,v) \ll N$ then there is a root α, say of $F_1(X,1) = 0$, such that $u - \alpha v \ll \sqrt{N} \ll N$. Similarly, since $f_1(u,v) \ll N$, we have $u - \beta v \ll N$ for some root β of $f_1(X,1)$. By subtraction we obtain $(\beta - \alpha)v \ll N$. Since F_1 and f_1 are coprime we will have $\alpha \neq \beta$, whence $v \ll N$. Similarly we have $u \ll N$. This provides the necessary bounds on u and v.

We have finally to consider the case in which F_1 is linear, so that $F(u,v) = c(au + bv)^k$, say. We then have an identity

$$f_1(u,v)^d + f_2(u,v)^d + f_3(u,v)^d = c(au + bv)^k,$$

and it is clear that d must divide k. But then Lemma 5 applies, since $d \geqslant 8$. Thus at least two of the terms must be proportional. A second application of the lemma then shows either that all four terms are proportional, contradicting the coprimality of f_1, f_2 and f_3, or that $f_i^d + f_j^d$ vanishes identically for some pair of indices $i \neq j$. In the latter case the corresponding solutions to

(6.9) cannot involve strictly positive integers. This completes the proof of Theorem 17.

Acknowledgements

These lecture notes were prepared while the author was visiting the Max-Planck Institute for Mathematics in Bonn. The hospitality and financial support of the Institute is gratefully acknowledged. Thanks are also due to Dr Browning and to Professors Perelli and Viola for their meticulous checking of these notes, the early versions of which contained numerous errors.

References

[1]   M. A. Bennett, N. P. Dummigan and T. D. Wooley, *The representation of integers by binary additive forms*, Compositio Math. 111 (1998), 15-33.

[2]   E. Bombieri and J. Pila, *The number of integral points on arcs and ovals*, Duke Math. J. 59 (1989), 337-357.

[3]   T. D. Browning, *Equal sums of two k-th powers*, J. Number Theory 96 (2002), 293-318.

[4]   N. D. Elkies, *On $A^4 + B^4 + C^4 = D^4$*, Math. Comp. 51 (1988), 825-835.

[5]   N. D. Elkies, *Rational points near curves and small nonzero $|x^3 - y^2|$ via lattice reduction*, Algorithmic number theory (Leiden, 2000), 33-63, Lecture Notes in Comput. Sci. 1838 (Springer, Berlin, 2000).

[6]   P. Erdős, *Arithmetical properties of polynomials*, J. London Math. Soc. 28 (1953), 416-425.

[7]   P. Erdős and K. Mahler, *On the number of integers that can be represented by a binary form*, J. London Math. Soc. 13 (1938), 134-139.

[8]   G. Faltings, *Endlichkeitssätze für abelsche Varietäten über Zahlkörpern*, Invent. Math. 73 (1983), 349-366.

[9]   M. L. Green, *Some Picard theorems for holomorphic maps to algebraic varieties*, Amer. J. Math. 97 (1975), 43-75.

[10] J. Harris, *Algebraic geometry. A first course*, Graduate Texts in Mathematics 133 (Springer-Verlag, New York, 1992).

[11] D. R. Heath-Brown, *Cubic forms in ten variables*, Proc. London Math. Soc. (3) 47 (1983), 225-257.

[12] D. R. Heath-Brown, *The density of rational points on non-singular hypersurfaces*, Proc. Indian Acad. Sci. (Math. Sci.) 104 (1994), 13-29.

[13] D. R. Heath-Brown, *The density of rational points on cubic surfaces*, Acta Arith. 79 (1997), 17-30.

[14] D. R. Heath-Brown, *Counting rational points on cubic surfaces*, Astérisque 251 (1998), 13-29.

[15] D. R. Heath-Brown, *The density of rational points on curves and surfaces*, Annals of Math. 155 (2002), 553-595.

[16] C. Hooley, *On binary cubic forms*, J. Reine Angew. Math. 226 (1967), 30-87.

[17] C. Hooley, *On the representations of a number as the sum of four cubes, I*, Proc. London Math. Soc. (3) 36 (1978), 117-140.

[18] C. Hooley, *On the numbers that are representable as the sum of two cubes*, J. Reine Angew. Math. 314 (1980), 146-173.

[19] C. Hooley, *On another sieve method and the numbers that are a sum of two h-th powers*, Proc. London Math. Soc. (3) 43 (1981), 73-109.

[20] C. Hooley, *On binary quartic forms*, J. Reine Angew. Math. 366 (1986), 32-52.

[21] C. Hooley, *On another sieve method and the numbers that are a sum of two h-th powers, II*, J. Reine Angew. Math. 475 (1996), 55-75.

[22] C. Hooley, *On binary cubic forms, II*, J. Reine Angew. Math. 521 (2000), 185-240.

[23] D. J. Lewis and K. Mahler, *On the representation of integers by binary forms*, Acta Arith. 6 (1960/61), 333-363.

[24] K. Mahler, *A note on hypothesis K of Hardy and Littlewood*, J. London Math. Soc. 11 (1936), 136-138.

[25] M. Nair, *Power free values of polynomials*, Mathematika 23 (1976), 159-183.

[26] D. J. Newman and M. Slater, *Waring's problem for the ring of polynomials*, J. Number Theory 11 (1979), 477-487.

[27] A. M. Ostrowski, *Über die Bedeutung der Theorie der konvexen Polyheder für formale Algebra*, Jahresber. Deutsche Math.-Verein. 30 (1921), 98-99.

[28] J. Pila, *Density of integral and rational points on varieties*, Astérisque 228 (1995), 183-187.

[29] W. M. Schmidt, *Integer points on hypersurfaces*, Monatsh. Math. 102 (1986), 27-58.

[30] C. M. Skinner and T. D. Wooley, *Sums of two k-th powers*, J. Reine Angew. Math. 462 (1995), 57-68.

Conversations on the Exceptional Character

Henryk Iwaniec

Department of Mathematics, Rutgers University
110 Frelinghuysen road, Piscataway, NJ 08854, USA
e-mail: iwaniec@math.rutgers.edu

1 Introduction

Everything has its exceptional character, and the analytic number theory is
no exception, it has one which is real and most perplexed. In this article I
will tell the story how the existence or the non-existence of such a character
shaped developments in arithmetic, especially for studies in the distribution of
prime numbers. Many researchers are affected by this dangerous yet beautiful
beast, and this author is no exception. I shall address questions and present
results which I witnessed during my own studies.

Of course, the Grand Riemann Hypothesis for the Dirichlet L-functions
rules out any exception! Nevertheless, after powerful researchers made serious
attacks on the beast and got painfully defeated, it is now understandable that
these people consider the problem to be as hard as the GRH itself. Some
experts go further with prediction that the GRH will be established first for
complex zeros, while the real zeros may wait long for a different treatment. In
the meantime we have many ways of living with or without the exceptional
character. In this article I try to show that this little dose of uncertainty is
enjoyable and stimulating for many new ideas.

Acknowledgements

I would like to thank the C.I.M.E. for supporting my participation in this
summer school in a wonderful scenery, and I am grateful to Alberto Perelli
and Carlo Viola for their encouragement, patience and great help in preparing
this article for publication.

Supported by the NSF grant DMS-0301168

2 The exceptional character and its zero

The characters $\chi \, (\mathrm{mod}\ D)$ were introduced by G. L. Dirichlet for his proof of the equidistribution of primes in reduced residue classes modulo D, the essential ingredient being the non-vanishing of the series

$$L(s,\chi) = \sum_1^\infty \chi(n)n^{-s} \tag{2.1}$$

at $s = 1$. It is already in this connection that the case of real character is different from all the complex characters.

Throughout we assume that $\chi = \chi_D$ is the real, primitive character of conductor D, so it is given by the Kronecker symbol

$$\chi(n) = \left(\frac{D^*}{n}\right), \tag{2.2}$$

where $D^* = \chi(-1)D$. This character is associated with the field

$$K = \mathbb{Q}(\sqrt{D^*}), \tag{2.3}$$

which is real quadratic if $\chi(-1) = 1$, or imaginary quadratic if $\chi(-1) = -1$. The celebrated Class Number Formula of Dirichlet asserts that

$$L(1,\chi) = \frac{\pi h}{\sqrt{D}} \quad \text{if } D^* < -4 \tag{2.4}$$

and similar formula holds in other cases. Here $h = h(-D)$ is the class number of K. By the obvious bound $h \geqslant 1$ one gets

$$L(1,\chi) \geqslant \frac{\pi}{\sqrt{D}}. \tag{2.5}$$

Hence $L(1,\chi) \neq 0$, but one can also show this directly as follows.

Consider the convolution $\lambda = 1 * \chi$, i.e.

$$\lambda(n) = \sum_{d|n} \chi(d) = \prod_{p^\alpha \| n} (1 + \chi(p) + \cdots + \chi(p^\alpha)) \geqslant 0.$$

For squares we have $\lambda(m^2) \geqslant 1$. Hence

$$T(x) = \sum_{n \leqslant x} \lambda(n)n^{-1/2} > \frac{1}{2} \log x.$$

On the other hand we find that

$$T(x) = \sum_{dm \leqslant x} \chi(d)(dm)^{-1/2}$$

$$= \sum_{d \leqslant y} \chi(d)d^{-1/2} \sum_{m \leqslant x/d} m^{-1/2} + \sum_{m \leqslant x/y} m^{-1/2} \sum_{y < d \leqslant x/m} \chi(d)d^{-1/2}$$

$$= \sum_{d < y} \frac{\chi(d)}{d^{1/2}} \left\{ 2\left(\frac{x}{d}\right)^{1/2} + c + O\left(\left(\frac{d}{x}\right)^{1/2}\right) \right\} + O\left(Dx^{1/2}y^{-1}\right)$$

$$= 2L(1,\chi)x^{1/2} + O\left(Dx^{1/2}y^{-1} + x^{-1/2}y\right)$$

$$= 2L(1,\chi)\sqrt{x} + O(\sqrt{D})$$

for $x \geqslant D$ by choosing $y = (xD)^{1/2}$, where the implied constant is absolute. Letting $x \to \infty$ these inequalities imply

$$L(1,\chi) \neq 0. \tag{2.6}$$

For showing (2.6) the class number formula (2.4) is dispensable, but it is a good starting place for estimating the class number $h = h(-D)$ of the imaginary quadratic fields. To this end one needs estimates of $L(1,\chi)$ (clearly (2.5) would give nothing new). By the Riemann Hypothesis for $L(s,\chi)$ it follows that

$$(\log \log D)^{-1} \ll L(1,\chi_D) \ll \log \log D, \tag{2.7}$$

hence the corresponding bounds for the class number

$$\sqrt{D}(\log \log D)^{-1} \ll h(D) \ll \sqrt{D} \log \log D. \tag{2.8}$$

Here the implied constants are absolute, effectively computable.

No chance to prove (2.8) by means available today. The best known upper bound is $L(1,\chi_D) \ll \log D$, which is easy (up to a constant). A lower bound for $L(1,\chi_D)$ is more important and the current knowledge is even less satisfactory. This problem is closely related to the zero-free region for $L(s,\chi_D)$.

At present we know that $L(s,\chi) \neq 0$ for $s = \sigma + it$ in the region

$$\sigma > 1 - \frac{c}{\log D(|t|+1)}, \tag{2.9}$$

where c is a positive absolute constant, for any character $\chi \pmod{D}$ with at most one exception. The exceptional character is real and the exceptional zero is real and simple. This follows by classical arguments of de la Vallée-Poussin (cf. E. Landau [L1]). Hence the question:

DOES THE EXCEPTIONAL ZERO EXIST?

In this article we shall try to illuminate this matter in bright and dark colors.

Let $\chi \pmod{D}$ be the real primitive character of conductor D and $\beta = \beta_\chi$ be the largest real zero of $L(s,\chi)$. Conjecturally $\beta_\chi = 0, -1$ if $\chi(-1) = 1, -1$, respectively. We say that χ is exceptional if

$$\beta > 1 - \frac{c}{\log D} \qquad (2.10)$$

for some positive constant c. One could make this concept more definite by fixing a sufficiently small value of the constant c, however we feel this would only obscure the presentation.

E. Landau [L1] said that H. Hecke knew that if χ was not exceptional, then

$$L(1, \chi) \gg (\log D)^{-1}. \qquad (2.11)$$

Remark. In the exceptional case of odd character $\chi = \chi_D$ (which is associated with the imaginary quadratic field $K = \mathbb{Q}(\sqrt{-D})$) there are several, quite precise relations between β_χ, h and $L(1, \chi)$, see [GSc], [G1], [GS].

Landau made a first breakthrough in the exceptional zeros area. Let $\chi \,(\mathrm{mod}\ D)$ and $\chi' \,(\mathrm{mod}\ D')$ be two distinct real primitive characters and β, β' be real zeros of $L(s, \chi)$, $L(s, \chi')$, respectively. He showed that

$$\min(\beta, \beta') \leqslant 1 - \frac{b}{\log DD'} \qquad (2.12)$$

with some positive, absolute constant b. This shows that the exceptional zeros occur very rarely. For example, calibrating the constant c in (2.10) to $c = b/3$ one can infer from (2.12) that if $\chi \,(\mathrm{mod}\ D)$ is exceptional then the next exceptional one $\chi' \,(\mathrm{mod}\ D')$ appears no sooner than for some $D' \geqslant D^2$.

There is a great idea in Landau's arguments which is still exercised in modern works. Generalising slightly we owe to Landau the product L-function (a quadratic lift)

$$\mathrm{Lan}(s, f) = L(s, f) L(s, f \otimes \chi) = \sum_1^\infty a_f(n) n^{-s} \qquad (2.13)$$

where

$$L(s, f) = \sum_1^\infty \lambda_f(n) n^{-s} \qquad (2.14)$$

can be any natural L-function and $L(s, f \otimes \chi)$ is derived from $L(s, f)$ by twisting (= multiplying) its coefficients $\lambda_f(n)$ with $\chi(n)$. This is particularly interesting for L-functions having finite degree Euler product. Then the prime coefficients of $\mathrm{Lan}(s, f)$ are

$$a_f(p) = \lambda(p)\lambda_f(p) \qquad (2.15)$$

where

$$\lambda(p) = 1 + \chi(p). \qquad (2.16)$$

The key observation is that if $L(1, \chi)$ is small then the class number h is small and $\chi(p) = 1$ is a rare event (not many primes split in the field K). Therefore $\chi(p) = -1$ and $a_f(p) = 0$ quite often. In other words $\chi(m)$ pretends to be the Möbius function $\mu(m)$ on squarefree numbers. In this scenario $L(s, f \otimes \chi)$ pretends to be $L(s, f)^{-1}$ up to a small Euler product, and $\mathrm{Lan}(s, f)$ behaves like a constant. This indicates that $L(s, f)$ cannot vanish at s near one, unless $L(s, f \otimes \chi)$ has a pole at $s = 1$ (natural L-functions are regular at $s \neq 1$).

Remark. Landau worked with

$$\zeta(s) L(s, \chi) L(s, \chi') L(s, \chi\chi')$$

which is the product of two $\mathrm{Lan}(s, f)$ for $\zeta(s)$ and $L(s, \chi)$.

Paraphrasing the above observation one may say that if the exceptional zero is very close to $s = 1$ then the other zeros are further away of $s = 1$; not only the zeros of $L(s, \chi)$, but also of any other L-function. This kind of a repelling property of the exceptional zero was nicely exploited in the works of M. Deuring [D] and H. Heilbronn [H] with a remarkable result that

$$h(-D) \to \infty \quad \text{as} \quad D \to \infty. \tag{2.17}$$

Shortly after that, E. Landau [L2] performed a quantitative analysis of the repelling effects and made a cute logical maneuver ending up with the lower bound

$$h(-D) \gg D^{\frac{1}{8}-\varepsilon} \tag{2.18}$$

for any $\varepsilon > 0$, the implied constant depending on ε (the original statement was a little different, but easily equivalent to (2.18)). In the same year and the same journal (the very first volume of Acta Arithmetica of 1936) C. L. Siegel [S] published the still stronger estimate

$$h(-D) \gg D^{\frac{1}{2}-\varepsilon}. \tag{2.19}$$

Note. Siegel was a much broader mathematician than Landau, however in my opinion Landau's ideas pioneered the above developments, so why did Siegel ignore Landau's contribution entirely?

3 How was the class number problem solved?

All three results (2.17), (2.18), (2.19) suffer from the serious defect of being ineffective (the implied constants in the Landau–Siegel estimates are not computable in terms of ε). For that reason one cannot use the results for the determination of all the imaginary quadratic fields with a given fixed class number h. Gauss conjectured that there are exactly nine fields with $h = 1$ (that is to say with unique factorization), the last one for $K = \mathbb{Q}(\sqrt{-163})$. Before the problem was completely solved it was known that there can be at most one more such field.

The Class Number One problem was eventually solved by arithmetical means (complex multiplication and Weber invariants) by K. Heegner [He] and later re-done independently by H. Stark [St]. A completely different solution was given by A. Baker [B] using transcendental number theory means (linear forms in three logarithms of algebraic numbers). Next it was recognized that the linear forms in two logarithms could do the job, so the 1948 work of A. O. Gelfond and Yu. V. Linnik [GL] was sufficient to resolve the non-existence of the tenth discriminant. H. Stark also settled the class number two problem.

Recently A. Granville and H. Stark [GS] showed a new inequality between the class number $h(-D)$ of $K = \mathbb{Q}(\sqrt{-D})$ and reduced quadratic forms (a, b, c) of discriminant $-D$, that is solutions of the equation

$$-D = b^2 - 4ac \tag{3.1}$$

in integers a, b, c with

$$-a < b \leqslant a < c, \quad \text{or} \quad 0 \leqslant b \leqslant a = c. \tag{3.2}$$

Note that the reduction condition (3.2) means that the root

$$z = \frac{-b + i\sqrt{D}}{2a} \tag{3.3}$$

is in the standard fundamental domain of the modular group $\Gamma = \mathrm{SL}_2(\mathbb{Z})$. They showed that

$$h(-D) \geqslant \left(\frac{\pi}{3} + o(1)\right) \frac{\sqrt{D}}{\log D} \sum_{(a,b,c)} a^{-1}. \tag{3.4}$$

Since the principal form with $a = 1$ is always there we get $h(-D) \gg \sqrt{D}/\log D$, and with some extra work one can deduce from (3.4) that $\chi = \chi_D$ is not exceptional. Fine, but the formula (3.4) of Granville–Stark is conditional, they need a uniform abc-conjecture for number fields, specifically for the Hilbert class field which is an extension of K of degree $h(-D)$! In spite of this criticism I strongly recommend this paper for learning a number of beautiful arguments.

A new excitement arose with the work of D. Goldfeld [G2] who succeeded in giving an effective lower bound

$$h(-D) \gg \prod_{p|D} \left(1 - \frac{2\sqrt{p}}{p+1}\right) \log D. \tag{3.5}$$

We shall give a brief sketch how this remarkable bound is derived. But first we point out some historical facts. In principle there is no reason to abandon the repelling property of an exceptional zero; one can still produce an effective result provided such an exceptional zero has a numerical value. OK, but believing in the Grand Riemann Hypothesis one cannot expect to find a real zero of any natural L-function which would be qualified to play a role of the repellent. A close analysis of Siegel's arguments reveals that any zero $\beta > \frac{1}{2}$ has some power of repelling; although not as strong as the zero near the point $s = 1$, yet sufficient for showing effectively that

$$h(-D) \gg D^{\beta - \frac{1}{2}} (\log D)^{-1}.$$

The only hope along such ideas is to use an L-function which vanishes at the central point $\beta = \frac{1}{2}$, at least this assumption does not contradict the GRH. Hence the first question is: does the central zero have an effect on the class number? In the remarkable paper by J. Friedlander [F] we find the answer: yes it does, and the impact depends on the order of the central zero! The second question is: how to find L-functions which do vanish at the central point? If $L(s, f)$ is self-dual and has the root number -1, that is the complete function $\Lambda(s, f)$ which includes the local factors at infinite places satisfies the functional equation

$$\Lambda(s, f) = -\Lambda(1 - s, f), \tag{3.6}$$

then, of course, $L(\frac{1}{2}, f) = 0$. Alas, no such function was known until J. V. Armitage [A] gave an example of an L-function of a number field (the Dirichlet L-functions $L(s, \chi)$ cannot vanish at $s - \frac{1}{2}$ by a folk conjecture). After this example Friedlander was able to apply his ideas giving an effective estimate for the class number of relative quadratic extensions. His work anticipated further research by Goldfeld.

A lot more possibilities were offered by elliptic curves. According to the Birch–Swinnerton-Dyer conjecture, the Hasse–Weil L-function of an elliptic curve $E/\mathbb{Q}$ vanishes at the central point to the order equal to the rank of the group of rational points. Goldfeld needed an L-function with central zero of order at least three. It is easy to point out a candidate as it is easy to construct an elliptic curve of rank $g = 3$, but proving that it is modular with the corresponding L-function vanishing to that order is a much harder problem. Ten years after Goldfeld's publication such an L-function was provided by B. Gross and D. Zagier [GZ], making the estimate (3.5) unconditional. Still, to make (3.5) practical (for example for the determination of all the imaginary quadratic fields $K = \mathbb{Q}(\sqrt{-D})$ with the class number $h = 3, 4, 5$, etc.) one needs a numerical value of the implied constant; so J. Oesterlé [O] refined Goldfeld's work and obtained a pretty neat estimate (3.5) with the implied constant $1/55$.

The best one can hope for to obtain along Goldfeld's arguments is $h(-D) \gg (\log D)^{g-2}$ when an L-function with the central zero of multiplicity g is employed. However there are popular problems which require a better effective lower bound for $h(-D)$, such as the

EULER IDONEAL NUMBER PROBLEM. Find all discriminants $-D$ for which the class group of $K = \mathbb{Q}(\sqrt{-D})$ has exactly one class in each genus.

By the genus theory, if $-D$ is an ideoneal discriminant then $h(-D) = 2^{\omega(D)-1}$, where $\omega(D)$ is the number of distinct prime divisors of D. Because

$\omega(D)$ can be as large as $\log D/\log\log D$, the problem of Euler calls for an effective lower bound

$$h(-D) \gg D^{c/\log\log D}, \quad \text{with} \ c > \log 2. \tag{3.7}$$

Of course, Landau's estimate (2.18) tells us that the number of idoneal discriminants is finite, yet we cannot determine all of them.

4 How and why do the central zeros work?

Very briefly we mention the main ideas behind the bound (3.5). There is no particular reason to restrict ourselves to the Hasse–Weil L-functions of elliptic curves, except that they are natural and available sources for multiple central zeros.

Let $f \in S_k(N)$ be a primitive cusp form of weight $k \geqslant 2$, k-even, and level N, that is a Hecke form on $\Gamma_0(N)$. This has the Fourier expansion

$$f(z) = \sum_1^\infty \lambda_f(n)n^{(k-1)/2}e(nz) \tag{4.1}$$

with coefficients $\lambda_f(n)$ which are eigenvalues of Hecke operators T_n for all n. With our normalization the associated L-function

$$L(s,f) = \sum_1^\infty \lambda_f(n)n^{-s} \tag{4.2}$$

converges absolutely in $\mathrm{Re}\, s > 1$ (because of the Ramanujan conjecture $|\lambda_f(n)| \leqslant \tau(n)$ proved by P. Deligne), it has the Euler product

$$L(s,f) = \prod_p \left(1 - \lambda_f(p)p^{-s} + \chi_0(p)p^{-2s}\right)^{-1} \tag{4.3}$$

where $\chi_0 \,(\mathrm{mod}\ N)$ is the principal character, and the complete product

$$\Lambda(s,f) = \left(\frac{\sqrt{N}}{2\pi}\right)^s \Gamma\left(s + \frac{k-1}{2}\right)L(s,f) \tag{4.4}$$

(which is entire) satisfies the self-dual functional equation

$$\Lambda(s,f) = w(f)\Lambda(1-s,\,f). \tag{4.5}$$

Here $w(f) = \pm1$ is called the root number, or the sign of the functional equation.

Let $\chi = \chi_D$ be the real character (the Kronecker symbol) associated with the imaginary quadratic field $K = \mathbb{Q}(\sqrt{-D})$. For simplicity assume that $(D,N) = 1$. The twisted form

$$f_\chi(z) = \sum_1^\infty \chi(n)\lambda_f(n)n^{(k-1)/2}e(nz) \tag{4.6}$$

is also a primitive form of weight k and level $N_\chi = ND^2$, the L-function

$$L(s, f_\chi) = \sum_1^\infty \chi(n)\lambda_f(n)n^{-s} \tag{4.7}$$

has appropriate Euler product, while the complete product

$$\Lambda(s, f_\chi) = \left(\frac{D\sqrt{N}}{2\pi}\right)^s \Gamma\left(s + \frac{k-1}{2}\right) L(s, f_\chi) \tag{4.8}$$

is entire and satisfies the functional equation

$$\Lambda(s, f_\chi) = w(f_\chi)\Lambda(1 - s, \, f_\chi) \tag{4.9}$$

with the root number

$$w(f_\chi) = \chi(-N)w(f). \tag{4.10}$$

Given f and χ we consider the Landau product

$$L(s) = L(s, f)L(s, f_\chi) = \sum_1^\infty a(n)n^{-s}. \tag{4.11}$$

This is an L-function with Euler product of degree four. The complete product

$$\Lambda(s) = Q^s \Gamma^2\left(s + \frac{k-1}{2}\right) L(s, f)L(s, f_\chi) \tag{4.12}$$

with

$$Q = \frac{DN}{4\pi^2} \tag{4.13}$$

satisfies the functional equation

$$\Lambda(s) = w\Lambda(1 - s), \qquad w = \chi(-N). \tag{4.14}$$

From here we compute the derivative of order $g \geqslant 0$ of $\Lambda(s)$ at $s = \frac{1}{2}$ by way of moving the integration in

$$\frac{g!}{2\pi i} \int_{(1)} \Lambda(s + \tfrac{1}{2})s^{-g-1}\mathrm{d}s.$$

We obtain

$$Q^{-1/2}\Lambda^{(g)}(1/2) = (1 + (-1)^g w)S \tag{4.15}$$

where

$$S = \sum_1^\infty \frac{a(n)}{\sqrt{n}} V\left(\frac{n}{Q}\right) \qquad (4.16)$$

and $V(y)$ is the Mellin transform of $g!\Gamma^2(s + k/2)s^{-g-1}$.

Assuming that $L(s)$ vanishes at $s = \frac{1}{2}$ of order larger than g and that

$$w = (-1)^g \qquad (4.17)$$

we get

$$S = 0. \qquad (4.18)$$

This is not possible if the class number $h = h(-D)$ is very small. The key point is that many coefficients $a(n)$ vanish so S is well approximated by a product over small primes. Waving hands a bit we can pull out from S a positive factor which takes squares, then we reduce S to a sum which looks like

$$S^\flat = \sum_{m<Q}^\flat \frac{a(m)}{\sqrt{m}} \left(\log \frac{Q}{m}\right)^g . \qquad (4.19)$$

Here the superscript $\flat$ means that the summation is restricted to squarefree numbers. For m squarefree we have

$$a(m) = \lambda(m)\lambda_f(m) \ll \lambda(m)\tau(m).$$

One also shows that

$$\sum_{y<m\leqslant x}^\flat \lambda(m)\tau(m)m^{-1/2} \ll h\left(\frac{1}{y} + \frac{x}{D}\right)^{1/2}. \qquad (4.20)$$

Now, another crucial point is that the sum $S^\flat$ runs over $m < Q$ with $Q \ll D$ (for N fixed), so (4.20) is extremely sharp in this range, giving

$$S^\flat = \sum_{m<y} \frac{a(m)}{\sqrt{m}} \left(\log \frac{Q}{m}\right)^g + O(h) \qquad (4.21)$$

with $y = (\log D)^{2g}$. Now assuming that $h \ll (\log D)^{g-2}$ one can approximate the short sum (4.21) essentially by the product

$$(\log D)^g \prod_{p<y} \left(1 + \frac{a(p)}{\sqrt{p}}\right) \qquad (4.22)$$

and eventually one draws a contradiction. Of course, in details the arguments are more complicated (cf. [IK]), but their key points look like above. As we mentioned in the transition from S to $S^\flat$ a factor taking squares is pulled out, this factor is essentially $L(1, \text{sym}^2 f)$ which is positive. We would also like to point out that the L-functions $L(s, f)$ for automorphic forms f on GL_n with $n > 2$ would not do the job, because the corresponding partial sum $S^\flat$ is longer than D. On the other hand if we could employ $L(s, f)$ with f on GL_1 then S^b is of length $\sqrt{D}$ and arguments similar to the above yield

$$h(-D) \gg D^{1/4} \log D. \tag{4.23}$$

Well, this is a wishful thinking; the GL_1 automorphic L-functions are just the Dirichlet L-functions for real characters, and none of these is expected to vanish at the central point $s = \frac{1}{2}$!

However one can derive the effective bound (4.23) from the more plausible hypothesis that

$$L(1/2, \chi_D) \geqslant 0. \tag{4.24}$$

Indeed we have (cf. (22.60) of [IK])

$$\zeta_K(1/2) = \zeta(1/2)L(1/2, \chi_D) = \frac{1}{2} \sum_{(a,b,c)} a^{-1/2} \log\left(\frac{\sqrt{D}}{2a}\right) + O(hD^{-1/4}),$$

where (a, b, c) runs over reduced forms, so $1 \leqslant a \leqslant \sqrt{D/3}$. Since $\zeta(\frac{1}{2})L(\frac{1}{2}, \chi_D) \leqslant 0$ this yields

$$h \gg \sum \left(\frac{\sqrt{D}}{a}\right)^{1/4} \log \frac{\sqrt{D}}{a} \tag{4.25}$$

giving (4.23) from just one term $a = 1$ (the principal form).

Because of the spectacular consequence (4.23) of the assumption (4.24), it seems that the latter is out of reach by the current technology. Of course, the GRH implies (4.24), but it also implies (2.8), so taking this road is pointless.

Closing this section we state an effective lower bound for $h(-D)$ which can be rigorously established by following the above guidelines.

Theorem 4.1 *Suppose that $L(s)$ given by (4.11) vanishes at $s = \frac{1}{2}$ to order $m \geqslant 3$. Then*

$$h(-D) \gg \theta(D) (\log D)^{g-1} \tag{4.26}$$

where $g = m - 1$ or $g = m - 2$ according to the parity condition $(-1)^g = w$. Here $\theta(D)$ is a mild factor, precisely

$$\theta(D) = \prod_{p|D} \left(1 + \frac{1}{p}\right)^{-3} \left(1 + \frac{2\sqrt{p}}{p+1}\right)^{-1}$$

while the implied constant depends only on the cusp form $f \in S_k(N)$ and is effectively computable.

Remark. For the purpose of proving Goldfeld's lower bound (3.5) Gross–Zagier delivered the following elliptic curve

$$E : \ -139\, y^2 = x^3 + 10\, x^2 - 20\, x + 8, \tag{4.27}$$

which is modular of conductor $N = 37 \cdot 139^2$ and rank $r = 3$.

5 What if the GRH holds except for real zeros?

If you are not afraid of confrontation with complex zeros of the $L(s)$, then be more productive working with the logarithmic derivative $L'(s)/L(s)$ rather than with the partial sums of $L^{(g)}(s)$. P. Sarnak and A. Zaharescu [SZ] have taken this route to improve Goldfeld's bound (3.5) significantly.

Theorem 5.1 Let $L(s, f)$ vanish at $s = \frac{1}{2}$ of order $\geqslant 3$. Let $-D$ be a fundamental discriminant with $\chi_D(N) = 1$. Suppose $L(s) = L(s, f)L(s, f \otimes \chi_D)$ has all its zeros either on the critical line $\operatorname{Re} s = \frac{1}{2}$ or on the real line $\operatorname{Im} s = 0$. Then

$$h(-D) \gg D^{\frac{1}{6}-\delta} \tag{5.1}$$

for any $\delta > 0$, the implied constant depending effectively on δ and f.

Theorem 5.1 is our variation on the work of Sarnak–Zaharescu. Their arguments are somewhat different and their bound (5.1) has the exponent $1/10$ in place of $1/6$. Moreover they worked only with the L-functions associated with the elliptic curve (4.27). But they also established a few other interesting results, some of which are ineffective.

To explain what is behind the proof of Theorem 5.1 we appeal to the so called "explicit formula"

$$\sum_{L(\rho)=0} \phi\Big(\frac{\gamma}{2\pi} \log R\Big) = 2\,\widehat{\phi}(0) \frac{\log D}{\log R} + \phi(0)$$

$$- 2\sum_{p \nmid D} \lambda_f(p) \frac{\lambda(p)}{\sqrt{p}} \,\widehat{\phi}\Big(\frac{\log p}{\log R}\Big) \frac{\log p}{\log R} + O\Big(\frac{\log \log D}{\log R}\Big). \tag{5.2}$$

This is derived by integrating $L'(s)/L(s)$ against a test function ϕ, using the functional equation (4.14) and Cauchy's residue theorem. Here $\phi(x)$ is an even function whose Fourier transform $\widehat{\phi}(y)$ is continuous and compactly supported, so $\phi(x)$ is entire, $R \geqslant 2$ is a parameter to be chosen later, and the implied constant depends only on the cusp form $f \in S_k(N)$ and the test function ϕ. To be fair we must admit that the exact explicit formula contains terms over prime powers which we put into the error term; this involves an estimate for the logarithmic derivative of $L(s, \operatorname{sym}^2 f)$ which follows by using the standard zero-free region near $s = 1$.

Suppose $\widehat{\phi}(y)$ is supported in $[-1, 1]$. Thinking of $h = h(-D)$ being small, say $h \leqslant D^{\frac{1}{6}-\delta}$, we can estimate the sum over primes in (5.2) by

$$\sum_{p < R} \frac{\lambda(p)}{\sqrt{p}} \frac{\log p}{\log R} \ll (\log D)^{-\delta} \tag{5.3}$$

for any R with $hD^{1/2} \leqslant R \leqslant h^{-2}D^{1-3\delta}$. Later we shall choose $R = D^{\frac{2}{3}-\delta}$. This is not an easy bound; it shows that $\lambda(p) = 1 + \chi(p) = 0$ very often. Hence the explicit formula (5.2) reduces to

$$\sum_{L(\rho)=0} \phi\!\left(\frac{\gamma}{2\pi}\log R\right) = 2\,\widehat{\phi}(0)\,\frac{\log D}{\log R} + \phi(0) + O\big((\log D)^{-\delta}\big). \qquad (5.4)$$

Now we are ready to play with (5.4), that is to say we want to pick up a test function $\phi(x)$ for which (5.4) is false. Already at first glance (5.4) is an improbable expression for most reasonable $\phi(x)$, because for what reason the zeros $\rho = \beta + i\gamma$ of $L(s)$ can be so regularly distributed to generate the functional $\phi \longrightarrow 3\,\widehat{\phi}(0) + \phi(0)$? As we do not know much about the spacing of zeros, our chance for contradiction goes by estimations. More chance if we can make every term $\phi((\gamma/2\pi)\log R)$ non-negative, so we can pick up the largest one and drop the others. For this reason we assume that all the zeros lay on two lines, $\beta = \frac{1}{2}$ or $\gamma \doteq 0$. Specifically we choose the Fourier pair (as in Sarnak–Zaharescu)

$$\phi(x) = \left(\frac{\sin \pi x}{\pi x}\right)^2, \qquad \widehat{\phi}(y) = \max(1 - |y|, 0) \qquad (5.5)$$

giving

$$m \leqslant 2\,\frac{\log D}{\log R} + 1 + O\big((\log D)^{-\delta}\big), \qquad (5.6)$$

where m is the multiplicity of the zero of $L(s)$ at $s = \frac{1}{2}$. For $R = D^{\frac{2}{3}-\delta}$ this implies $m < 4$, that is $m \leqslant 3$. However we assumed that $L(s, f)$ has zero at $s = \frac{1}{2}$ of order $\geqslant 3$, and we also know that $L(s) = L(s, f)L(s, f \otimes \chi)$ has the root number $w = \chi(-N) = \chi(-1) = -1$, so m is even, $m \geqslant 4$. This contradiction completes the proof of Theorem 5.1.

Remarks. The final blow in the proof of Theorem 5.1 is powered by the positivity arguments. This is an excellent example of the strength of the real-variable harmonic analysis when coupled with the positivity ideas. The positivity arguments are hard to implement to complex domains, so the hypothesis that all zeros are on specific lines is critical.

6 Subnormal gaps between critical zeros

A simple central zero of an L-function yields no effect on the class number, still if it has large order then it does. But what about the complex zeros on the critical line, so to speak the critical zeros, which appear in abundance? More hopefully one should ask if some clustering of the critical zeros can be as effective as the high order central zero. This possibility was contemplated in the literature long before the central zero effects. Indeed the fundamental work of H. L. Montgomery [M] on the pair correlation of zeros was motivated by the class number problems. In a joint paper Montgomery–Weinberger [MW] used zeros of a fixed real Dirichlet L-function which are close to the central point $s = \frac{1}{2}$, by means of which they were able to perform quite strong computations for the imaginary quadratic fields $K = \mathbb{Q}(\sqrt{-D})$ with the class number $h = h(-D) = 1, 2$ (see [MW] for precise results and for some other relevant claims).

Recently B. Conrey and H. Iwaniec [CI] considered the Hecke L-function

$$L(s; \psi) = \sum_{\mathfrak{a}} \psi(\mathfrak{a})(N\mathfrak{a})^{-s} \tag{6.1}$$

associated with the imaginary quadratic field $K = \mathbb{Q}(\sqrt{-D})$. Here $\mathfrak{a}$ runs over the non-zero integral ideals of K and $\psi \in \widehat{\mathcal{Cl}}(K)$ is a character of the class group. Although $L(s; \psi)$ does not factor as the Landau product (2.13) (unless ψ is a genus character), it possesses the same crucial feature, namely the lacunarity of the coefficients

$$\lambda_\psi(n) = \sum_{N\mathfrak{a}=n} \psi(\mathfrak{a}) \tag{6.2}$$

if the class number $h(-D)$ is ridiculously small.

Have in mind that the corresponding theta series

$$\theta(z; \psi) = \sum_0^\infty \lambda_\psi(n) \, e(nz),$$

with $\lambda_\psi(0) = h/2$ for the trivial character and $\lambda_\psi(0) = 0$ otherwise, is a modular form of weight $k = 1$, level D and Nebentypus χ_D (it is a cusp form if ψ is a complex character). This yields the functional equation (self-dual)

$$\Lambda(s; \psi) = \left(\frac{\sqrt{D}}{2\pi}\right)^s \Gamma(s) \, L(s; \psi) = \Lambda(1 - s; \psi). \tag{6.3}$$

By contour integration one can show that the number of zeros of $L(s; \psi)$ in the rectangle $s = \sigma + it$ with $0 \leqslant \sigma \leqslant 1$, $0 < t \leqslant T$ satisfies

$$N(T; \psi) = \frac{T}{\pi} \log \frac{T\sqrt{D}}{2\pi e} + O(\log DT). \tag{6.4}$$

Hence one can say (assuming GRH) that the average gap between consecutive zeros $\rho = \frac{1}{2} + i\gamma$ and $\rho' = \frac{1}{2} + i\gamma'$ is about $\pi/\log\gamma$.

We have shown in [CI] that if the gap is a little smaller than the average for sufficiently many pairs of zeros on the critical line (no GRH is required) then $h(-D) \gg \sqrt{D}(\log D)^{-A}$ for some constant $A > 0$. This result may not appeal to everybody, because our L-function $L(s; \psi)$ is intimately related with the field $K = \mathbb{Q}(\sqrt{-D})$, so are its zeros. Well, we can draw a more impressive statement from the zeta function of K (the case of the trivial class group character)

$$\zeta_K(s) = \zeta(s)L(s, \chi_D). \tag{6.5}$$

Since we do not need all the zeros, we choose only those of $\zeta(s)$ which apparently have nothing in common with the character χ_D.

Theorem 6.1 *Let $\rho = \frac{1}{2} + i\gamma$ denote the zeros of $\zeta(s)$ on the critical line and $\rho' = \frac{1}{2} + i\gamma'$ denote the nearest zero to ρ on the critical line ($\rho' = \rho$ if ρ is a multiple zero). Suppose*

$$\#\left\{\rho \; ; \; 0 < \gamma < T, \; |\gamma - \gamma'| \leqslant \frac{\pi}{\log \gamma}\left(1 - \frac{1}{\sqrt{\log \gamma}}\right)\right\} \gg T(\log T)^{4/5} \qquad (6.6)$$

for any $T \geqslant 2005$. Then

$$h(-D) \gg \sqrt{D}\,(\log D)^{-90} \qquad (6.7)$$

where the implied constant is effectively computable.

Have in mind that each of $\zeta(s)$, $L(s, \chi_D)$ has asymptotically half the number of zeros of $\zeta_K(s)$, so that relative to $\zeta(s)$ in Theorem 6.1 we are counting the gaps which are a little smaller than the half of the average gap. Our condition (6.6) is quite realistic, because the Pair Correlation Conjecture of Montgomery asserts that the zeros of $\zeta(s)$ are not equidistributed. In fact the PCC implies that

$$|\gamma - \gamma'| < \frac{2\pi\vartheta}{\log \gamma} \qquad (6.8)$$

with any $\vartheta > 0$, for a positive proportion of zeros. The best unconditional estimate (6.8) is known with $\vartheta = 0.68$ by Montgomery–Odlyzko [MO], $\vartheta = 0.5171$ by Conrey–Ghosh–Gonek [CGG] and $\vartheta = 0.5169$ by Conrey–Iwaniec (unpublished). For the effective bound (6.7) we need (6.8) with some $\vartheta < \frac{1}{2}$.

Remark. At the meeting in Seattle of August 1996 D. R. Heath-Brown gave a talk "Small Class Number and the Pair Correlation of Zeros" in which he showed how the assumption of the class number being small distorts the Pair Correlation Conjecture of Montgomery. His and our arguments have similar roots.

The main principles of the proof of Theorem 6.1 can be seen quickly (but of course, the details are formidable) from the "approximate functional equation"

$$L(s; \psi) = \sum_{n \leqslant t\sqrt{D}} \lambda_\psi(n)\, n^{-s} + X(s) \sum_{n \leqslant t\sqrt{D}} \lambda_\psi(n)\, n^{s-1} + \dots$$

on the line $s = \frac{1}{2} + it$. Because $\lambda_\psi(n)$ are lacunary (assuming the class number is relatively small) the two partial sums can be shortened substantially, so the variation of $L(s; \psi)$ in t is mostly controlled by the gamma factor

$$X(\tfrac{1}{2} + it) = \left(\frac{2\pi e}{t\sqrt{D}}\right)^{2it}\{1 + O(1/t)\}$$

(a "root number" in the t-aspect). In other words the "infinite place" leads the spin while the "finite places" are too weak and too few to disturb. Therefore in this illusory scenario the zeros of $L(s; \psi)$ should follow the equidistribution law, but we postulated otherwise, hence the contradiction.

From the above discussion one may also get an idea why the PCC predicts a density function for differences between zeros to be other than constant; the reason might be that the "finite places" generate the periodicities n^{it} with distinct frequencies as n varies around t.

Another interesting lesson one can draw from the above situation is that the very popular perception that the zeros of very different L-functions operate in their own independent ways, that they do not see each other so cannot conspire, is not wise. This idealistic view may appeal to math philosophers, but when the tools of analytic number theory break the sky we find a more fascinating and complex structure.

7 Fifty percent is not enough!

... for winning in a democracy, neither for ruling out the exceptional character. In recent investigations we (see Iwaniec–Sarnak [IS]) took an opposite direction for attacking the problem of the exceptional character. Rather than using the central zeros of L-functions as repellents, we need families of L-functions whose central values are positive, not very small.

For this presentation we take the set $\mathcal{H}_k(N)$ of cusp forms f of weight $k \geqslant 2$, k-even which are primitive on the group $\Gamma_0(N)$ (i.e. which are eigenfunctions of all the Hecke operators T_n, $n = 1, 2, 3, \ldots$). The basic properties of the associated L-functions are (4.1)–(4.14). The Hilbert space structure of the linear space $S_k(N)$ plays a role in our arguments (the Petersson formula brings Kloosterman sums which are our tools), and the transition from spectral to arithmetical normalizations is achieved by the factors

$$\omega_f = \zeta_N(2) \, L(1, \mathrm{sym}^2 f)^{-1}, \tag{7.1}$$

where $\zeta_N(s)$ denotes the zeta function with the local factors at primes $p \mid N$ omitted, and $L(s, \mathrm{sym}^2 f)$ is the L-function associated with the symmetric square representation of f. These are mild factors since

$$(\log kN)^{-2} \ll L(1, \mathrm{sym}^2 f) \ll (\log kN)^2. \tag{7.2}$$

The upper bound is an easy consequence of the Ramanujan conjecture (proved by P. Deligne), while the lower bound is essentially saying that $L(s, \mathrm{sym}^2 f)$ has no exceptional zero which is now known as fact due to Hoffstein–Lockhart [HL]. Actually we do not make use of (7.2), because the factors ω_f are kept present in our averagings over the family $\mathcal{H}_k(N)$. We have

$$\sum_{f \in \mathcal{H}_k(N)} \omega_f X_f \sim N \tag{7.3}$$

for each of the vectors $X_f = 1$, $X_f = L(\frac{1}{2}, f)$, $X_f = L(\frac{1}{2}, f_\chi)$, and

$$\sum_{f \in \mathcal{H}_k(N)} \omega_f L(\tfrac{1}{2}, f) L(\tfrac{1}{2}, f_\chi) \sim N L(1, \chi) \tag{7.4}$$

as $N \to \infty$ over squarefree numbers, uniformly for $D \leqslant N^\delta$ with $\delta > 0$ a small fixed constant (recall that D is the conductor of $\chi = \chi_D$).

The great attraction of the formula (7.4) is the fact that the central values $L(\tfrac{1}{2}, f)$, $L(\tfrac{1}{2}, f_\chi)$ are known unconditionally to be non-negative. Of course, one can deduce this from the GRH, yet we can do it without (see Waldspurger [Wa], Kohnen–Zagier [KZ], Katok–Sarnak [KS], Guo [Gu]). The non-negativity of $L(\tfrac{1}{2}, f)$ has much to do with f being a GL$_2$ form. Recall that this property for self-dual GL$_1$ forms (the Dirichlet real characters) would have immediate consequences for the class number (see (4.23)), unfortunately it is not provable without recourse to the GRH. It is not easy to show that $L(\tfrac{1}{2}, f) \geqslant 0$, $L(\tfrac{1}{2}, f_\chi) \geqslant 0$ for any cusp form $f \in \mathcal{H}_k(N)$, but these estimates are not actually deep. One may get an idea why the central values are non-negative by considering a simple example of f whose coefficients are

$$a(n) = \sum_{ad=n} \chi(a) \overline{\chi}(d) (a/d)^{ir}.$$

In this case

$$L(\tfrac{1}{2}, f) = \left| L(\tfrac{1}{2} + ir, \chi) \right|^2 \geqslant 0.$$

One may express the central values of automorphic L-functions by sums of squares in a more profound fashion. For the CM forms F. Rodríguez Villegas [R-V] takes squares of a theta-series. A cute proof of $L(\tfrac{1}{2}, f) \geqslant 0$ follows as a by-product in the recent investigations of W. Luo–P. Sarnak [LS] in quantum chaos.

Another important feature of the asymptotic formula (7.4) is its "purity", that is to say the absence of lower order terms involving the derivative $L'(1, \chi)$. Therefore if $L(1, \chi)$ is very small then almost all the products $L(\tfrac{1}{2}, f) L(\tfrac{1}{2}, f_\chi)$ are very small. Before speculating further let us restrict the summation (7.4) to forms for which the root number of $L(s, f) L(s, f_\chi)$ is one,

$$w = w_f w_{f_\chi} = w_f w_f \chi(-N) = \chi(-N) = 1$$

(because the L-functions with root number -1 vanish at the central point trivially by the functional equation).

One can establish that a lot of $L(\tfrac{1}{2}, f)$ and $L(\tfrac{1}{2}, f_\chi)$ are not very small, say

$$L(\tfrac{1}{2}, f) \geqslant (\log N)^{-2} \tag{7.5}$$

$$L(\tfrac{1}{2}, f_\chi) \geqslant (\log N)^{-2}, \tag{7.6}$$

using the classical idea of averaging of mollified values. If the two sets of f's for which both (7.5) and (7.6) hold had a large intersection (positive percentage) we could conclude from (7.3), (7.4) by the non-negativity that

$$L(1,\chi) \gg (\log D)^{-4}.$$ (7.7)

We did succeed to show that (7.5) holds for at least 50% of forms $f \in \mathcal{H}_k(N)$ with $\varepsilon_f = 1$, and that (7.6) holds for at least 50% of forms $f \in \mathcal{H}_k(N)$ with $\varepsilon_{f_\chi} = 1$. These results are just too short to ensure a significantly large intersection.

It is hard to believe that a character $\chi \pmod{D}$ can be so vicious to divide (by twisting) any respectful family of L-functions into two equal size classes (almost), giving all the power to one class and nothing for the other class. Yet, we cannot destroy such feature by present tools. Having (7.6) for 50% forms it suffices to get (7.5) for slightly more than 50%. The latter task seems to be quite promising, because the character issue is irrelevant! Not really! Actually we undertook the task with stronger tools offered by averaging over the level N. Consequently we were able to attach to $L(\frac{1}{2}, f)$ a mollifying factor longer than N (which puts us beyond diagonal) leading to (7.5) for more than 50% of the forms f with $w_f = 1$. But (7.6) is not useful for every N, here we need the root number condition $\chi(-N) = 1$. Ironically, if one installs this condition to averaging over the level, then the off-diagonal terms are badly affected, and the excess over 50% disappears! We are convinced there is a magic conspiracy out there which prevents us from cracking the existence of the exceptional character along our lines.

Perhaps one should build a comprehensive theory which explains all the peculiar loops in which we are often trapped when venturing beyond the diagonal path.

8 Exceptional primes

An easy way of handling problems is to avoid them. Better yet one may find that the obstacle which is hard to eliminate can be exploited to reach the goal in other ways. The case of the exceptional character is a spectacular example in this regard. We shall present a few applications of the exceptional character for producing primes in tide areas where even the GRH fails to work. Having tasted the results one may only wish that the exceptional character is a real thing, not an illusion which researchers of several generations tried to kill.

The good reason for liking the real exceptional characters $\chi(m)$ is that they pretend to be the Möbius function $\mu(m)$ at almost all squarefree integers m. In the same time the characters are periodic functions, so one can apply a Fourier analysis in place of zeros of L-functions. One needs a quantitative measure of how closely $\chi(m)$ approximates to $\mu(m)$. To this end consider

$$\Delta(z,x) = \sum_{z < n \leqslant x} \lambda(n) n^{-1}.$$ (8.1)

Recall that $\lambda = 1 * \chi$, and χ is the real character of conductor D, not necessarily exceptional. We have

$$\Delta(z, x) = L(1, \chi) \left[\log x + O(\log z)\right] \tag{8.2}$$

if $x > z \geqslant D^2$. Hence $\lambda(n)$ vanishes very often if $L(1, \chi)$ is very small, and $\chi(p) = -1$ very often. We can see this phenomenon better from estimates for

$$\delta(z, x) = \sum_{z \leqslant p < x} \lambda(p) p^{-1}. \tag{8.3}$$

By the inequality $\delta(z, x) \Delta(1, z) \leqslant \Delta(z, xz)$ we get by (8.2)

$$\delta(z, x) \Delta(1, z) \leqslant L(1, \chi) \left[\log x + O(\log z)\right]. \tag{8.4}$$

Applying the trivial bound $\Delta(1, z) \geqslant 1$ we get

$$\delta(z, x) \leqslant L(1, \chi) \left[\log x + O(\log z)\right] \tag{8.5}$$

if $x > z \geqslant D^2$. One can also estimate $\delta(z, x)$ in terms of any real zero, say β, of $L(s, \chi)$. Indeed we have

$$\begin{aligned}
\Delta(1, z) &> z^{\beta-1} \sum_{1 \leqslant n < z} \lambda(n) n^{-\beta} \left(1 - \frac{n}{z}\right) \\
&= \frac{1}{2\pi i} \int_{(1)} \zeta(s + \beta) L(s + \beta, \chi) z^{s+\beta-1} \frac{ds}{s(s+1)} \\
&= L(1, \chi)(1 - \beta)^{-1}(2 - \beta)^{-1} + O(q^{1/4} z^{-1/2}) \\
&> L(1, \chi)(1 - \beta)^{-1}
\end{aligned}$$

by moving the integration to the line $\operatorname{Re} s = \frac{1}{2} - \beta$, provided $x > z \geqslant D^2$. Inserting this bound to (8.4) we obtain

$$\delta(z, x) < (1 - \beta) \left[\log x + O(\log z)\right]. \tag{8.6}$$

The implied constants in (8.5), (8.6) are absolute. These inequalities show that $\delta(z, x)$ is very small so $\chi(p) = -1$ for almost all p in the range $D^2 \leqslant p \leqslant D^A$, A constant, provided χ is exceptional.

Now how this observation can be used for applications to prime numbers? We start from the zeta-function of $K = \mathbb{Q}(\sqrt{D^*})$

$$\zeta_K(s) = \sum_1^\infty \lambda(n) n^{-s} = \zeta(s) L(s, \chi). \tag{8.7}$$

Define the multiplicative function $\nu(m)$ by

$$\frac{1}{\zeta_K(s)} = \sum_1^\infty \nu(m) m^{-s}. \tag{8.8}$$

Note that $|\nu(m)| \leqslant \lambda(m)$ and

$$\nu(m) = \mu(m)\lambda(m) \quad \text{if } m \text{ is squarefree.} \tag{8.9}$$

If $\lambda(n)$ is lacunary (i.e. $\lambda(n)$ vanishes very often), then so is $\mu(m)$. Next, writing $\zeta'(s)/\zeta(s) = L(s,\chi)\zeta'(s)/\zeta_K(s)$ we find

$$\Lambda(n) = \sum_{klm=n} \chi(k)(\log l)\nu(m). \tag{8.10}$$

We also introduce the function

$$\lambda'(d) = \sum_{kl=d} \chi(k)\log l, \tag{8.11}$$

so (8.10) becomes

$$\Lambda(n) = \sum_{dm=n} \lambda'(d)\nu(m). \tag{8.12}$$

One can easily view $\lambda'(d)$ as a divisor-like function, because log is smooth and slowly increasing while χ is periodic with a relatively small period. Moreover, if χ is exceptional then $\nu(m)$ is lacunary, so it contributes to (8.12) very little only at small m. Therefore one can see (8.12) as an approximation to the von Mangoldt function by a divisor-like function. By means of this formula in many interesting applications one can accomplish results for primes as strong as for the divisor function.

The situation described above is a little bit oversimplified. In practice a serious difficulty occurs with handling the lacunary part of (8.12), say

$$\Lambda_*(n) = \sum_{\substack{dm=n \\ m>D^2}} \lambda'(d)\nu(m), \tag{8.13}$$

especially when $\Lambda_*(n)$ is applied against a sparse sequence $\mathcal{A} = (a_n)$. We estimate (8.13) by

$$|\Lambda_*(n)| \leqslant \tau(n)(\log n) \sum_{\substack{m|n \\ m>D^2}} \lambda(m).$$

We deal with $\tau(n)\log n$ crudely by special devices which allow us to ignore this factor, so we are left essentially with

$$\Lambda_\infty(n) = \sum_{\substack{m|n \\ m>D^2}} \lambda(m). \tag{8.14}$$

From the above partitions we arrive at (essentially)

$$\sum_{n\leqslant x} a_n\Lambda(n) = \sum_{\substack{dm\leqslant x \\ m\leqslant D^2}} a_{dm}\lambda'(d)\nu(m) + O\left((\log x)^{2005} \sum_{n\leqslant x} a_n\Lambda_\infty(n)\right).$$

In estimating the sum

$$\sum_{n \leqslant x} a_n \Lambda_\infty(n) \tag{8.15}$$

we cannot forget that in our mind $\lambda(m)$ is lacunary. If $\mathcal{A} = (a_n)$ is not sparse then one can disconnect a_n from $\Lambda_\infty(n)$ by Cauchy's inequality and estimate the resulting sums separately and quite easily. A great challenge appears for very sparse sequences. We open the convolution $\lambda = 1 * \chi$ in (8.14) and consider $\Lambda_\infty(n)$ to be like the divisor function $\tau_3(n)$ rather than like $\tau(n)$ in the main term. After having opened the $\lambda(m)$ our analysis of the error term must be asymptotically accurate, because at the end we must observe a crucial cancellation which reflects the lacunarity of $\lambda(m)$.

Having said this we conclude that the existence of the exceptional character creates a useful substitute for $\Lambda(n)$ in terms of divisor-like functions of degree three. Therefore various methods of analytic number theory which are capable of showing an asymptotic formula for

$$\sum_{n \leqslant x} a_n \tau_3(n) \tag{8.16}$$

are likely to be modified to yield an asymptotic formula for

$$\sum_{n \leqslant x} a_n \Lambda(n). \tag{8.17}$$

In a series of papers J. Friedlander and H. Iwaniec [FI2], [FI3], [FI4] realized these ideas for a few very sparse sequences. For example we got the following formula for primes in a short interval

$$\psi(x) - \psi(x - y) = y\left\{1 + O\left(L(1, \chi)(\log x)^{r^r}\right)\right\} \tag{8.18}$$

for $x \geqslant y \geqslant x^{39/79}$, $x \geqslant D^r$ where $r = 18{,}290$ and the implied constant is absolute. The result is unconditional, but it is useful only under special conditions, such as

$$L(1, \chi) \leqslant (\log D)^{-1 - r^r}$$

and $D^r \leqslant x \leqslant D^{2r}$. Note that $\frac{39}{79} < \frac{1}{2}$, so the interval in (8.18) can be very short. The Riemann Hypothesis does not work for intervals shorter than $[x - \sqrt{x}, \ x]$.

Similar ideas (however more precise with respect to the powers of logarithms) were used earlier by D. R. Heath-Brown [H-B1] with an impressive conclusion that if there is an infinite sequence of exceptional zeros, then there are infinitely many twin primes.

9 The least prime in an arithmetic progression

9.1 Introduction

In the previous sections we have been trying either to eliminate the exceptional character from the surface of the Earth, or to employ it for producing impressive, yet illusory results. However one can play both tunes in a complementary fashion to end up with completely unconditional results and effective ones, unlike the Landau–Siegel type. The celebrated work of Yu. V. Linnik [L] on the least prime in an arithmetic progression is a true masterpiece of this kind.

Let $p_{\min}(q, a)$ denote the first prime $p \equiv a \pmod q$. Linnik proved that

$$p_{\min}(q, a) \ll q^L \qquad (9.1)$$

for any $q > 1$, $(a, q) = 1$, where L and the implied constant are absolute and effectively computable. The GRH gives (9.1) with any $L > 2$, while the best known result with $L = 5.5$ is due to D. R. Heath-Brown [H-B3]. The best possible (9.1) should be with any $L > 1$.

Using arguments similar to these in the proof of (8.18) we [FI2] showed that

$$\psi(x; q, a) = \frac{\psi(x)}{\varphi(q)} \left\{ 1 + O\big(L(1, \chi)(\log x)^{r^r}\big) \right\} \qquad (9.2)$$

for $(a, q) = 1$, $D \nmid q$ and any $x \geqslant \max\left\{ q^{462/233}, D^r \right\}$ with $r = 554, 401$, where the implied constant is absolute. If $\chi \pmod D$ is exceptional in the sense that

$$L(1, \chi) \leqslant (\log D)^{-1-r^r} \qquad (9.3)$$

then (9.2) implies (9.1) with $L = 2 - \frac{1}{59}$ for q in the range

$$D^r \leqslant q \leqslant \exp\big(L(1, \chi)^{-r/(r^r+1)}\big). \qquad (9.4)$$

Earlier Heath-Brown [H-B2] also succeeded in bringing the Linnik constant L close to 2, but not below 2, under the assumption of the existence of exceptional characters (our condition (9.3) is a bit stronger).

As we described in the previous section at some point our arguments depend on the sum (8.16), specifically for the sequence $\mathcal{A} = (a_n)$ of the characteristic function of the progression $n \equiv a \pmod q$. By no means this is an easy sum if $x < q^2$; just mention we had to modify the result of [FI1], which is proved by an appeal to the Riemann Hypothesis for varieties.

Back to Linnik's bound (9.1) there are several interesting points to say about its original proof in regard to the theory of Dirichlet L-functions. All the proofs up to now use essentially the following three principles:

P1: THE ZERO-FREE REGION (2.9).

P2: THE LOG-FREE ZERO DENSITY ESTIMATE:

$$\sum_{\chi \,(\mathrm{mod}\, q)} N_\chi(\alpha, T) \leqslant a(qT)^{b(1-\alpha)} \qquad (9.5)$$

where $N_\chi(\alpha, T)$ denotes the number of zeros $\rho = \beta + i\gamma$ of $L(s, \chi)$ with $\beta \geqslant \alpha$, $|\gamma| < T$ for $\frac{1}{2} \leqslant \alpha \leqslant 1$, $T \geqslant 1$, and a, b are positive absolute constants.

P3: The Exceptional Zero Repulsion:

If $\beta > 1 - c/\log q$ is a real zero of $L(s,\chi)$ for a real character $\chi \pmod{q}$, then there is no other zero of any L-functions with characters modulo q in the region

$$\sigma \geqslant 1 - \frac{d \mid \log(1-\beta)\log q \mid}{\log q(|t|+1)} \tag{9.6}$$

where c, d are positive, small, absolute constants.

The first principle is classical, the other two are due to Linnik. The principles P2, P3 set the theory of Dirichlet L-functions at the most profound level. Yes, they will be obsolete soon after the GRH is proved, but for the time being (perhaps a very long time) these principles are treasures on their own right.

Having paid tribute to P2, P3, we are going to show the Linnik bound (9.1) without using these principles. Our arguments (a joint work with J. Friedlander) reveal a new potential of sieve methods. First we treat the case when the exceptional character is available, because the arguments are quick and require almost nothing from the theory of L-functions, not even P1 nor the Prime Number Theorem. The second case, with no exceptional character existing, is somewhat longer. In this case we do use P1, however by a hard work one could dispense with it. The point is that using sieve we are not aiming at an asymptotic formula for primes $p \equiv a \pmod{q}$, so the "parity barrier" of linear sieve is not a problem, the primes can be produced along the elementary lines à la Chebyshev. Anyway, there is no reason to work hard without P1, when the derivation of the zero-free region (2.9) is by today standards very easy.

We replace P2 by a much simpler result:

Proposition 9.1 *Let $\rho = \beta + i\gamma$ run over the zeros of $L(s,\chi)$ for a character $\chi \pmod{q}$. Put*

$$A(t) = \sum_{\rho} \left(1 + (1-\beta)\log q\right)^{-1}\left(1 + |t - \gamma|\log q\right)^{-2}. \tag{9.7}$$

For any real t we have

$$A(t) \leqslant \frac{3}{2} + \frac{\log(|t| + c)}{2 \log q} \tag{9.8}$$

where c is a positive absolute constant.

Remark. A bound for $A(t)$ with $|t| \leqslant q$ by any fixed number suffices for our applications, because we are not going to give a numerical value of the constant L in (9.1).

Proof of (9.8). For any $s = \sigma + it$ with $1 < \sigma \leqslant 2$ we have

$$-\operatorname{Re}\frac{L'}{L}(s,\chi) = \frac{1}{2}\log q|s| - \sum_\rho \operatorname{Re}\frac{1}{s-\rho} + O(1),$$

$$\left|\frac{L'}{L}(s,\chi)\right| \leqslant -\frac{\zeta'}{\zeta}(\sigma) = \frac{1}{\sigma-1} + O(1),$$

$$\operatorname{Re}\frac{1}{s-\rho} = \frac{\sigma-\beta}{(\sigma-\beta)^2+(t-\gamma)^2} \geqslant \frac{1}{\sigma-\beta}\left(1+\frac{|t-\gamma|}{\sigma-1}\right)^{-2}.$$

Hence

$$\sum_\rho \frac{\sigma-1}{\sigma-\beta}\left(1+\frac{|t-\gamma|}{\sigma-1}\right)^{-2} \leqslant 1 + \frac{1}{2}(\sigma-1)\left(\log q|s| + O(1)\right).$$

For $\sigma = 1+1/\log q$ the left side is equal to $A(t)$ giving the bound (9.8).

9.2 The case with an exceptional character

Let $\chi \pmod q$ be a real, non-principal character. We do not really assume that χ is exceptional, so we end up with unconditional results, which will be useful in the final conclusion only when χ is the exceptional character.

We shall apply sieve to the sequence $\mathcal{A} = (\lambda(n)a_n)$, where $\lambda = 1 * \chi$ and a_n is the characteristic function of the arithmetic progression $n \equiv a \pmod q$ with $(a,q)=1$. Clearly we must assume that

$$\chi(a) = 1, \tag{9.9}$$

or else there is nothing but zero in $\mathcal{A}$. We need to evaluate the sums of type

$$A_d(x) = \sum_{\substack{n\leqslant x \\ n\equiv 0\,(d)}} \lambda(n)a_n = \sum_{\substack{n\leqslant x/d \\ n\equiv a\overline{d}\,(q)}} \lambda(dn)$$

for $(d,q)=1$. Think of $\lambda(n)$ as the Hecke eigenvalues of the Eisenstein series of weight one and the central character (Nebentypus) χ. Hence $\lambda(n)$ is multiplicative,

$$\lambda(dn) = \sum_{\delta|(d,n)} \mu(\delta)\chi(\delta)\lambda(d/\delta)\lambda(n/\delta).$$

This yields

$$A_d(x) = \sum_{\delta|d} \mu(\delta)\chi(\delta)\lambda(d/\delta)A(x/\delta d; q, a\overline{\delta d})$$

where

$$A(y;q,\alpha) = \sum_{\substack{m\leqslant y \\ m\equiv\alpha\,(q)}} \lambda(m).$$

Here we write

$$\lambda(m) = \sum_{kl=m} \chi(k) = (1 + \chi(m)) \sum_{\substack{kl=m \\ k<l}} \chi(k) + \chi(\sqrt{m})$$

for $(m, q) = 1$, where the last term $\chi(\sqrt{m})$ vanishes, unless $\sqrt{m}$ is an integer (a traditional convention), that is if $m = k^2$. This gives

$$A(y; q, \alpha) = (1 + \chi(\alpha)) A^*(y; q, \alpha) + O(\sqrt{y})$$

where

$$A^*(y; q, \alpha) = \sum_{\substack{kl \leqslant y, \, k<l \\ kl \equiv \alpha \, (\mathrm{mod} \, q)}} \chi(k) = \frac{1}{q} \sum_{k<\sqrt{y}} \chi(k)\left(\frac{y}{k} - k\right) + O(\sqrt{y})$$

$$= L(1, \chi) \, y \, q^{-1} + O(\sqrt{y}).$$

Hence

$$A(y; q, \alpha) = (1 + \chi(\alpha)) L(1, \chi) \, y \, q^{-1} + O(\sqrt{y})$$

where the implied constant is absolute. Here we have $\alpha = a \, \overline{\delta} d$, $\chi(\alpha) = \chi(a)\chi(d/\delta) = \chi(d/\delta) = 1$, or else $\lambda(d/\delta) = 0$. Hence we obtain

$$A_d(x) = 2L(1, \chi) \frac{\nu(d)x}{dq} + O\left(\tau_3(d) \sqrt{x/d}\right) \tag{9.10}$$

where

$$\nu(d) = \sum_{\delta | d} \mu(\delta) \frac{\chi(\delta)}{\delta} \lambda\left(\frac{d}{\delta}\right).$$

This is multiplicative with

$$\nu(p) = 1 + \chi(p)\left(1 - \frac{1}{p}\right). \tag{9.11}$$

We write the approximation (9.10) in the familiar sieve format

$$A_d(x) = g(d)X + r_d(x) \tag{9.12}$$

where $g(d) = \nu(d)/d$ stands for the sifting density function,

$$X = 2L(1, \chi) \, x \, q^{-1} \tag{9.13}$$

and $r_d(x)$ is the error term, $r_d(x) \ll \tau_3(d)\sqrt{x/d}$. Hence the remainder term of level y satisfies

$$R(x, y) = \sum_{d<y} |r_d(x)| \ll \sqrt{xy} \, (\log y)^2,$$

where the implied constant is absolute.

We seek primes so we wish to estimate $S(\mathcal{A}, \sqrt{x})$. Under normal conditions the task would be beyond the capability of a linear sieve. However we think that $\chi(p) = -1$ very often for the exceptional character, so the density function at such primes is very small, $g(p) = p^{-2}$. In this scenario we have a sieve problem of small dimension, and the Fundamental Lemma of sieve theory does the job,

$$S(\mathcal{A}, z) = X V(z)\{1 + O(e^{-s})\} + O(\sqrt{xy}\,(\log y)^2) \qquad (9.14)$$

where $s = \log y/\log z \geqslant 2$ and

$$V(z) = \prod_{\substack{p<z \\ p\nmid q}} (1 - g(p)) = \prod_{\substack{p<z \\ p\nmid q}} \left(1 - \frac{1}{p}\right)\left(1 - \frac{\chi(p)}{p}\right). \qquad (9.15)$$

We do not need the full strength of e^{-s} in (9.14), a weaker term s^{-1} suffices. Choosing

$$y = \frac{x}{q^3(\log x)^8}, \qquad x \geqslant q^8, \qquad (9.16)$$

we see that the error term in (9.14) is negligible giving

$$S(\mathcal{A}, z) = X V(z)\left\{1 + O\left(\frac{\log z}{\log x}\right)\right\}. \qquad (9.17)$$

From $S(\mathcal{A}, z)$ we go to $S(\mathcal{A}, \sqrt{x})$ by Buchstab's formula

$$S(\mathcal{A}, \sqrt{x}) = S(\mathcal{A}, z) - \sum_{z \leqslant p < \sqrt{x}} S(\mathcal{A}_p, p).$$

For every $z \leqslant p < \sqrt{x}$ we estimate $S(\mathcal{A}_p, p)$ by an upper-bound sieve of level y/p getting

$$S(\mathcal{A}_p, p) \ll g(p)V(p)X + \sqrt{xy/p}\,(\log y)^2.$$

Adding these estimates we arrive at

$$S(\mathcal{A}, \sqrt{x}) = X V(z)\left\{1 + O\left(\frac{\log z}{\log x} + \delta(z, x)\right)\right\} \qquad (9.18)$$

where $\delta(z, x)$ is defined by (8.3) and was estimated twice in (8.5) and (8.6), in the range $x > z \geqslant q^2$. Hence we conclude (still unconditional result)

Lemma 9.2 *Let $\chi \pmod{q}$ be a real, non-principal character and β be any real zero of $L(s, \chi)$. Suppose $\chi(a) = 1$. Then for $x \geqslant q^8$ we have*

$$\pi(x; q, a) = 2L(1, \chi) V(q^2)\frac{x}{q}\left\{1 + O\left(\frac{\log q}{\log x} + (1 - \beta)\log x\right)\right\} \qquad (9.19)$$

where the implied constant is absolute. The factor $1 - \beta$ can be replaced by $L(1, \chi)$.

Corollary 9.3 *Let the condition of Lemma 9.2 be satisfied. Then for x in the segment*

$$q^A \leqslant x \leqslant e^{1/A(1-\beta)}, \qquad (9.20)$$

where A is any large constant, $A \geqslant 8$, we have

$$\pi(x; q, a) > \frac{x \, L^*(1, \chi)}{\varphi(q) \log q} \qquad (9.21)$$

where

$$L^*(1, \chi) = L(1, \chi) \prod_{p < q^2} \big(1 - \chi(p)\big). \qquad (9.22)$$

Of course, the segment (9.20) is not void only if $1 - \beta \leqslant A^{-2}(\log q)^{-1}$, which with a large constant A means that χ is an exceptional character. Assuming that this is the case we get the Linnik bound (9.1) with $L = A$.

9.3 A parity-preserving sieve inequality

Next we are going to apply sieve to the sequence $\mathcal{A} = (a_n)$ which is the characteristic function of the arithmetic progression $n \equiv a \pmod q$. Our goal is to estimate

$$S(\mathcal{A}, z) = \sum_{\substack{n \leqslant x \\ (n, P(z)) = 1}} a_n$$

for $z = \sqrt{x}$. For $(d, q) = 1$ we have

$$A_d(x) = \frac{x}{dq} + O(1)$$

so we have a problem of linear sieve. For the level of distribution of $\mathcal{A}$ we take

$$y = \frac{x}{q} (\log x)^{-4/3}. \qquad (9.23)$$

Recall that the lower-bound linear sieve works only in the range $z \leqslant \sqrt{y}$, which is not a problem because

$$S(\mathcal{A}, \sqrt{x}) = S(\mathcal{A}, \sqrt{y}) + O\left(\frac{x \log q}{\varphi(q)(\log x)^2}\right) \qquad (9.24)$$

by the Brun-Titchmarsh estimate.

Next the linear sieve gives (see [I], Rutgers notes)

$$S(\mathcal{A}, \sqrt{y}) = S^-(\mathcal{A}, \sqrt{y}) + \sum_{n \text{ even}} S_n(\mathcal{A}, \sqrt{y}).$$

Usually one discards all terms $S_n(\mathcal{A}, \sqrt{y})$ getting the lower bound

$$S(\mathcal{A}, z) \geqslant S^-(\mathcal{A}, z) = X V^-(y, z) + O(y),$$

where $X = xq^{-1}$ and $O(y)$ is the bound for the remainder term. The main term equals

$$V^-(y, z) = \{f(s) + O((\log y)^{-1/3})\} V(z)$$

with $s = \log y / \log z$. For $z = \sqrt{y}$ we get $f(2) = 0$, so the sum $S^-(\mathcal{A}, \sqrt{y})$ is negligible, and we must keep the terms $S_n(\mathcal{A}, \sqrt{y})$. We get

$$S(\mathcal{A}, \sqrt{y}) \geqslant \sum_{n \text{ even}} S_n(\mathcal{A}, \sqrt{y}) + O(y). \tag{9.25}$$

We only exploit the term for $n = 4$, which is

$$S_4(\mathcal{A}, \sqrt{y}) = \sum_{p_4 < p_3 < p_2 < p_1 < \sqrt{y}}' \cdots \sum_{}' S(\mathcal{A}_{p_1 p_2 p_3 p_4}, p_4)$$

where the summation is restricted by the conditions $p_1 p_2^3 < y$, $p_1 p_2 p_3 p_4^3 \geqslant y$. Dropping more terms we deduce that

$$S(\mathcal{A}, \sqrt{y}) \geqslant \frac{1}{24} \sum_{\substack{p_4 p_3 p_2 p_1 p \leqslant x \\ p_4 p_3 p_2 p_1 p \equiv a \, (q)}}'' \cdots \sum_{}'' \sum 1 + O(y) \tag{9.26}$$

where the superscript $''$ indicates that the prime variables p_r run independently over the segment

$$y^{1/6} < p_r < y^{1/5}, \qquad r = 1, 2, 3, 4. \tag{9.27}$$

Remarks. In (9.26) we have estimated a sum over primes (essentially) by a sum over products of five primes. The other sums $S_n(\mathcal{A}, \sqrt{y})$ with n even (which we discarded) run essentially over products of $n + 1$ primes (if y is close to x), so the parity is odd throughout all terms of (9.25). Therefore the formula (9.25) does not break the parity which is the barrier for getting primes within the traditional axioms of sieve theory.

We have chosen to work with products of five primes $p_4 p_3 p_2 p_1 p$ rather than three for technical advantage (products of larger fixed odd number of primes would also be fine).

Before applying characters to detect the congruence $p_4 p_3 p_2 p_1 p \equiv a \,(\mathrm{mod}\, q)$ we exploit the positivity, and partition the sum $\sum'' \cdots \sum'' \sum$ in (9.26) into suitable blocks so the separation of variables will not be an issue later. It is essential that we can do it at this point without much loss, because the forthcoming arguments will be so delicate that anything like partial summation will inflict unreparable damage (certainly losing a logarithmic factor will kill the arguments). To this end we fix a smooth function $f(u)$ supported on $\left[\frac{1}{2}, 1\right]$ with $0 \leqslant f(u) \leqslant 1$, and put

$$\psi_{\mathcal{X}}(x; q, a) = \sum \cdots \sum_{\substack{p\, p_1 p_2 p_3 p_4 \equiv a\ (q) \\ x_r < p_r < 2x_r}} f(p\, p_1 p_2 x_3 x_4 / x) \log p$$

where $\mathcal{X} = [x_1, x_2, x_3, x_4]$ runs over the vectors of dyadic partition points of the segment $[y^{1/6},\, y^{1/5}]$. Then (9.26) yields

$$S(\mathcal{A}, \sqrt{y}) \geqslant \frac{1}{24 \log x} \sum_{\mathcal{X}} \psi_{\mathcal{X}}(x; q, a) + O(y). \tag{9.28}$$

Notice that we did not partition p and we included the variables p_1, p_2 together with p in the argument of the smoothing function f, while p_3, p_4 are excluded from f. These seemingly technical devices will play nicely in relevant character sums.

Let $\psi_{\mathcal{X}}(x)$ denote the corresponding sum over $p\, p_1 p_2 p_3 p_4$ with the congruence condition dropped, i.e. $\psi_{\mathcal{X}}(x) = \psi_{\mathcal{X}}(x; 1, 1)$, so we have

$$\psi_{\mathcal{X}}(x) \asymp x/(\log x_1)(\log x_2)(\log x_3)(\log x_4), \tag{9.29}$$

where the implied constant depends only on f. Our goal is to show that

$$\psi_{\mathcal{X}}(x; q, a) \asymp \frac{\psi_{\mathcal{X}}(x)}{\varphi(q)} \tag{9.30}$$

for all relevant $\mathcal{X}$ subject to some conditions on x, q, a to be specified later. Hence we derive by (9.24), (9.28) that

$$\pi(x; q, a) \gg \frac{\pi(x)}{\varphi(q)} \tag{9.31}$$

subject to the same conditions on x, q, a.

9.4 Estimation of $\psi_{\mathcal{X}}(x; q, a)$

Applying the orthogonality of characters we write

$$\psi_{\mathcal{X}}(x; q, a) = \frac{1}{\varphi(q)} \sum_{\chi\ (\mathrm{mod}\ q)} \overline{\chi}(a)\, \psi_{\mathcal{X}}(x, \chi)$$

where

$$\psi_{\mathcal{X}}(x, \chi) = \sum \cdots \sum_{\substack{p\, p_1 p_2 p_3 p_4 \\ x_r < p_r < 2x_r}} \chi(p\, p_1 p_2 p_3 p_4)\, f(p\, p_1 p_2 x_3 x_4 / x) \log p. \tag{9.32}$$

Denote $\mathcal{W} = [x_1, x_2]$, $w = x/x_3 x_4$ and

$$\psi_W(w,\chi) = \sum_{\substack{p\,p_1p_2 \\ x_r<p_r<2x_r}} \chi(p\,p_1p_2)\,f(p\,p_1p_2/w)\log p, \qquad (9.33)$$

so (9.32) becomes

$$\psi_X(x,\chi) = \psi_W(w,\chi)\Big(\sum_{x_3<p_3<2x_3}\chi(p_3)\Big)\Big(\sum_{x_4<p_4<2x_4}\chi(p_4)\Big). \qquad (9.34)$$

The principal character $\chi_0 \,(\mathrm{mod}\,q)$ gives the main term. We also put aside the contribution of the real character, say $\chi_1 \,(\mathrm{mod}\,q)$, because it will require a special treatment when χ_1 is exceptional. We get

$$\psi_X(x;q,a) = \frac{1}{\varphi(q)}\left\{\psi_X(x) + \chi_1(a)\,\psi_X(x,\chi_1) + \Delta_X(x;q,a)\right\} \qquad (9.35)$$

where $\Delta_X(x;q,a)$ denotes the contribution of all the characters $\chi \neq \chi_0,\chi_1$.

We estimate $\Delta_X(x;q,a)$ in the following fashion which resembles the circle method for ternary additive problems (we have here a multiplicative analog):

$$\left|\Delta_X(x;q,a)\right| \leqslant$$
$$\max_{\chi\neq\chi_0,\chi_1}\left|\psi_W(w,\chi)\right|\Big(\sum_\chi\Big|\sum_{p_3}\chi(p_3)\Big|^2\Big)^{1/2}\Big(\sum_\chi\Big|\sum_{p_4}\chi(p_4)\Big|^2\Big)^{1/2}.$$

It is easy to see that for any $X \geqslant q^2$

$$\sum_{\chi\,(\mathrm{mod}\,q)}\Big|\sum_{X<p<2X}\chi(p)\Big|^2 \ll \left(\frac{X}{\log X}\right)^2$$

where the implied constant is absolute. To this end square out and estimate the resulting sum over primes $p_1 \equiv p_2 \pmod{q}$ by the Brun-Titchmarsh theorem.

Now we need a modest, but non-trivial estimate of $\psi_W(w,\chi)$ for every $\chi \neq \chi_0,\chi_1$ (it is like asking for a non-trivial estimate of the corresponding exponential sum in the circle method at every point of the minor arc). In Section 9.6 we prove that for $\chi \neq \chi_0,\chi_1$

$$\psi_W(w,\chi) \ll \left(\frac{\log q}{\log w} + \frac{1}{\log q}\right)\frac{w}{(\log x_1)(\log x_2)} \qquad (9.36)$$

provided $x_1, x_2 \geqslant q$ and $x_1x_2q^2 \leqslant w^{3/4}$. Hence we get

$$\Delta_X(x;q,a) \ll \left(\frac{\log q}{\log x} + \frac{1}{\log q}\right)\psi_X(x) \qquad (9.37)$$

where the implied constant is absolute (we assume that $x \geqslant q^8$).

For estimating $\psi_\mathcal{X}(x, \chi_1)$ with the real character χ_1 we have two options. First in Section 9.6 we prove that

$$\psi_\mathcal{W}(w, \chi_1) \ll \left(\frac{1}{(1 - \beta_1) \log w} + \frac{\log q}{\log w} + \frac{1}{\log q} \right) \frac{w}{(\log x_1)(\log x_2)}, \quad (9.38)$$

where β_1 is the largest real zero of $L(s, \chi_1)$. Hence by trivial estimations of sums over p_3, p_4 in (9.34) we get

$$\psi_\mathcal{X}(x, \chi_1) \ll \left(\frac{1}{(1 - \beta_1) \log x} + \frac{\log q}{\log x} + \frac{1}{\log q} \right) \psi_\mathcal{X}(x). \quad (9.39)$$

The second option is to replace every $\chi(p)$, $\chi(p_r)$, $r = 1, 2, 3, 4$ in (9.32) by -1 getting

$$\psi_\mathcal{X}(x, \chi_1) = -\{1 + O(\delta(z, x))\} \psi_\mathcal{X}(x)$$

where $\delta(z, x)$ is defined by (8.3). Using (8.6) we get

$$\psi_\mathcal{X}(x, \chi_1) = -\{1 + O((1 - \beta_1) \log x)\} \psi_\mathcal{X}(x). \quad (9.40)$$

9.5 Conclusion

We are now ready to derive Linnik's bound (9.1) from the assorted results in Sections 9.2, 9.3, 9.4.

Suppose $\chi_1 \pmod q$ is a non-principal real character such that $L(s, \chi_1)$ has a real zero β_1 with

$$(1 - \beta_1) \log q \leqslant A^{-2}, \quad (9.41)$$

where A is a large constant, $A \geqslant 8$. If $\chi_1(a) = 1$ then (9.21) yields (9.1) with $L = A$. If $\chi_1(a) = -1$ then (9.35), (9.40), (9.37) yield

$$\psi_\mathcal{X}(x; q, a) = \frac{2}{\varphi(q)} \psi_\mathcal{X}(x)\{1 + O(1/A)\}$$

for $q^A \leqslant x \leqslant e^{1/A(1-\beta_1)}$. Hence we get (9.31) which yields (9.1) with $L = A$.

Now we can assume that the largest real zero β_1 of $L(s, \chi_1)$ does not satisfy (9.41). Then we get by (9.35), (9.39), (9.37) that

$$\psi_\mathcal{X}(x; q, a) = \frac{1}{\varphi(q)} \psi_\mathcal{X}(x)\left\{1 + O\left(A^2 \frac{\log q}{\log x} + \frac{1}{\log q}\right)\right\}$$

$$\gg \frac{\psi_\mathcal{X}(x)}{\varphi(q)}$$

if $x \geqslant q^{A^2 B}$, where B is a large constant. Hence we get (9.31) for every $(a, q) = 1$, which yields (9.1) with $L = A^2 B$.

9.6 Appendix. Character sums over triple-primes

In this section we give a non-trivial estimate for the character sum $\psi_W(w, \chi)$ defined by (9.33).

Proposition 9.4 *Let* $\chi \pmod{q}$ *be a non-trivial character. Put*

$$\delta_\chi = \min_\rho \{1 - \beta \; ; \; |\gamma| \leqslant \log q\} \qquad (9.42)$$

where $\rho = \beta + i\gamma$ *denote zeros of* $L(s, \chi)$. *For* $x_1, x_2 \geqslant q$ *and* $x_1 x_2 q^2 \leqslant w^{3/4}$
we have

$$\psi_W(w, \chi) \ll \left\{ \frac{1}{\delta_\chi \log w} + \frac{1}{\log q} \right\} \frac{w}{(\log x_1)(\log x_2)} \qquad (9.43)$$

where the implied constant is absolute.

Clearly Proposition 9.4 and the classical zero-free region (2.9) imply (9.36) and (9.38).

Estimating trivially one gets $\psi_W(w, \chi) \ll w/(\log x_1)(\log x_2)$, so $\{\dots\}$ is the saving factor (if w, q are large).

The proof of Proposition 9.4 does not require the zero-free region, although at some point we use the Prime Number Theorem in the form

$$\psi(y) = y + O\big(y(\log y)^{-4}\big) \qquad (9.44)$$

which helps to simplify the arguments. We start by the "explicit formula"

$$\sum_n \chi(n) \Lambda(n) f(n/w) = - \sum_{\beta \geqslant 1/2} \widehat{f}(\rho) w^\rho + O\big(\sqrt{w}\,(\log q)^2\big),$$

where $\widehat{f}(s)$ is the Mellin transform of $f(u)$, $\rho = \beta + i\gamma$ run over the zeros of $L(s, \chi)$ and the implied constant is absolute. This gives

$$\psi_W(w, \chi) = - \sum_{\beta \geqslant 1/2} \widehat{f}(\rho) w^\rho \Big(\sum_{p_1} \chi(p_1) p_1^{-\rho} \Big) \Big(\sum_{p_2} \chi(p_2) p_2^{-\rho} \Big)$$
$$+ O\big(\sqrt{w}\,(\log q)^2\big).$$

Note that the error term absorbs the contribution of prime powers p^2, $p^3, \dots$ which are missing in $\psi_W(w, \chi)$. We have

$$\widehat{f}(\rho) \ll |\rho|^{-3}. \qquad (9.45)$$

Hence

$$\sum_{|\gamma| > T} |\widehat{f}(\rho)| \ll T^{-2} \log q. \qquad (9.46)$$

We choose $T = \log q$ and estimate the term $\psi_W(w, \chi)$ with $|\gamma| > T$ trivially, getting

$$\psi_W(w,\chi) = -\sum{}^{\rho} \hat{f}(\rho)\left(\frac{w}{x_1 x_2}\right)^{\rho}\left(\sum_{p_1} \chi(p_1)(x_1/p_1)^{\rho}\right)\left(\sum_{p_2} \chi(p_2)(x_2/p_2)^{\rho}\right)$$
$$+ O\big(w/(\log x_1)(\log x_2)\log q\big).$$

Here $\sum^{\rho}$ denotes summation of the zeros $\rho = \beta + i\gamma$ of $L(s,\chi)$ restricted by $\beta \geqslant \frac{1}{2}$, $|\gamma| \leqslant \log q$. We have

$$\left|\left(\frac{w}{x_1 x_2}\right)^{\rho}\right| \leqslant \frac{w}{x_1 x_2}\left(\frac{x_1 x_2 q^2}{w}\right)^{\delta_\chi} q^{2(\beta-1)},$$

$$\left(\frac{x_1 x_2 q^2}{w}\right)^{\delta_\chi} \leqslant w^{-\delta_\chi/4} \leqslant 4/\delta_\chi \log w.$$

Hence

$$\psi_W(w,\chi) \ll T_\chi(x_1,x_2)\, w/x_1 x_2 \delta_\chi \log w + w/(\log x_1)(\log x_2)\log q,$$

where

$$T_\chi(x_1,x_2) =$$
$$\sum{}^{\rho} q^{2(\beta-1)}\left|\sum_{x_1 < p_1 < 2x_1} \chi(p_1)(x_1/p_1)^{\rho}\right|\left|\sum_{x_2 < p_2 < 2x_2} \chi(p_2)(x_2/p_2)^{\rho}\right|. \quad (9.47)$$

Note that we ignored the factor (9.45) because it does not help, the problem occurs with bounded zeros.

To complete the proof of (9.43) it remains to show that

$$T_\chi(x_1,x_2) \ll x_1 x_2/(\log x_1)(\log x_2) \quad (9.48)$$

where the implied constant is absolute. If we knew that

$$\sum{}^{\rho} q^{2(\beta-1)} \ll 1, \quad (9.49)$$

then (9.48) would quickly follow by trivial estimation of the sums over primes p_1, p_2. The bound (9.49) is true, it is a kind of log-free density bound for the zeros of $L(s,\chi)$ of height $|\gamma| \leqslant \log q$. However we avoid (9.49) (whose proof would be quite long) by gaining a bit from cancellation in the sums over p_1, p_2. When the variation of $\rho = \beta + i\gamma$ with respect to γ exceeds $(\log q)^{-1}$ we do have a change in the argument of $(x_r/p_r)^{\rho}$ as p_r varies in $x_r < p_r < 2x_r$, $\log x_r \gg \log q$. This observation should explain why we did not want to separate p from p_1, p_2 in the smoothing function $f(p\,p_1 p_2/w)$. Moreover it is worth mentioning that for this purpose we use two prime variables p_1, p_2 rather than one, because we can apply the duality principle.

Lemma 9.5 *For $x \geqslant q$ and any complex numbers c_p we have*

$$\sum^{\rho} q^{2(\beta-1)} \Big| \sum_{x<p<2x} c_p(x/p)^{\rho} \Big|^2 \ll \frac{x}{\log x} \sum_p |c_p|^2 \qquad (9.50)$$

where the implied constant is absolute.

Clearly (9.48) follows from (9.50) by Cauchy's inequality.

For the proof of Lemma 9.5 it suffices to show that for any complex numbers a_{ρ}

$$\sum_{x<p<2x} \frac{\log p}{p} \Big| \sum^{\rho} a_{\rho}(x/p)^{\rho} q^{\beta-1} \Big|^2 \ll \sum^{\rho} |a_{\rho}|^2. \qquad (9.51)$$

First we smooth the outer summation by introducing a factor $h(p/x)$, then we square out and execute the summation over p getting

$$\sum_p \frac{\log p}{p} h(p/x)(x/p)^{\rho_1 + \bar{\rho}_2} \ll \big(1 + |\gamma_1 - \gamma_2| \log x\big)^{-2}.$$

This follows by partial summation using (9.44) and that the Fourier transform of $h(u)$ satisfies $\widehat{h}(v) \ll (1 + |v|)^{-2}$. Hence the left side of (9.51) is estimated by

$$\sum^{\rho_1} \sum^{\rho_2} |a_{\rho_1} a_{\rho_2}| \big(1 + |\gamma_1 - \gamma_2| \log q\big)^{-2} q^{\beta_1 + \beta_2 - 2} \leqslant \sum^{\rho} A(\gamma) |a_{\rho}|^2$$

where $A(t)$ is defined and estimated in Proposition 9.1. This proves (9.51), hence (9.50) by the duality, and finally (9.43).

References

[A]   J. V. Armitage, *Zeta functions with a zero at $s = \frac{1}{2}$*, Invent. Math. 15 (1972), 199-205.

[B]   A. Baker, *Linear forms in the logarithms of algebraic numbers*, Mathematika 13 (1966), 204-216.

[CGG]   J. B. Conrey, A. Ghosh and S. M. Gonek, *A note on gaps between zeros of the zeta function*, Bull. London Math. Soc. 16 (1984), 421-424.

[CI]   B. Conrey and H. Iwaniec, *Spacing of zeros of Hecke L-functions and the class number problem*, Acta Arith. 103 (2002), 259-312.

[D]   M. Deuring, *Imaginär-quadratische Zahlkörper mit der Klassenzahl (1)*, Math. Z. 37 (1933), 405-415.

[F]   J. B. Friedlander, *On the class numbers of certain quadratic extensions*, Acta Arith. 28 (1976), 391-393.

[FI1]   J. B. Friedlander and H. Iwaniec, *Incomplete Kloosterman sums and a divisor problem*, Ann. Math. (2) 121 (1985), 319-350.

[FI2]   J. B. Friedlander and H. Iwaniec, *Exceptional characters and prime numbers in arithmetic progressions*, Int. Math. Res. Notices 37 (2003), 2033-2050.

[FI3]   J. B. Friedlander and H. Iwaniec, *Exceptional characters and prime numbers in short intervals*, Selecta Math. 10 (2004), 61-69.

[FI4]   J. B. Friedlander and H. Iwaniec, *The illusory sieve*, preprint.

[GL]   A. O. Gelfond and Yu. V. Linnik, *On Thue's method and the effectiveness problem in quadratic fields*, Dokl. Akad. Nauk SSSR 61 (1948), 773-776 (in Russian).

[G1]   D. Goldfeld, *An asymptotic formula relating the Siegel zero and the class number of quadratic fields*, Ann. Scuola Norm. Sup. Pisa (4) 2 (1975), 611-615.

[G2]   D. Goldfeld, *The class number of quadratic fields and the conjectures of Birch and Swinnerton-Dyer*, Ann. Scuola Norm. Sup. Pisa (4) 3 (1976), 623-663.

[GSc]   D. Goldfeld and A. Schinzel, *On Siegel's zero*, Ann. Scuola Norm. Sup. Pisa (4) 2 (1975), 571-583.

[GS]   A. Granville and H. M. Stark, *ABC implies no "Siegel zeros" for L-functions of characters with negative discriminant*, Invent. Math. 139 (2000), 509-523.

[GZ]   B. Gross and D. Zagier, *Heegner points and derivatives of L-series*, Invent. Math. 84 (1986), 225-320.

[Gu]   J. Guo, *On the positivity of the central critical values of automorphic L-functions for* GL(2), Duke Math. J. 83 (1996), 157-190.

[H-B1]   D. R. Heath-Brown, *Prime twins and Siegel zeros*, Proc. London Math. Soc. (3) 47 (1983), 193-224.

[H-B2]   D. R. Heath-Brown, *Siegel zeros and the least prime in an arithmetic progression*, Quart. J. Math. Oxford (2) 41 (1990), no. 164, 405-418.

[H-B3]   D. R. Heath-Brown, *Zero-free regions for Dirichlet L-functions, and the least prime in an arithmetic progression*, Proc. London Math. Soc. (3) 64 (1992), 265-338.

[He]   K. Heegner, *Diophantische Analysis und Modulfunktionen*, Math. Z. 56 (1952), 227-253.

[H]   H. Heilbronn, *On the class-number in imaginary quadratic fields*, Quart. J. Math. Oxford 5 (1934), 150-160.

[HL]   J. Hoffstein and P. Lockhart, *Coefficients of Maass forms and the Siegel zero*, Ann. Math. (2) 140 (1994), 161-181.

[I]   H. Iwaniec, *Sieve Methods*, Graduate Course Notes, Rutgers, 1996.

[IK]   H. Iwaniec and E. Kowalski, *Analytic Number Theory*, Amer. Math. Soc. Colloquium Publications, vol. 53, 2004.

[IS]   H. Iwaniec and P. Sarnak, *The non-vanishing of central values of automorphic L-functions and Landau-Siegel zeros*, Israel J. Math. 120 (2000), 155-177.

[KS]   S. Katok and P. Sarnak, *Heegner points, cycles and Maass forms*, Israel J. Math. 84 (1993), 193-227.

[KZ]   W. Kohnen and D. Zagier, *Values of L-series of modular forms at the center of the critical strip*, Invent. Math. 64 (1981), 175-198.

[L1]   E. Landau, *Über die Klassenzahl imaginär-quadratischer Zahlkörper*, Gött. Nachr. (1918), 285-295.

[L2]   E. Landau, *Bemerkungen zum Heilbronnschen Satz*, Acta Arith. 1 (1936), 1-18.

[L]   Yu. V. Linnik, *On the least prime in an arithmetic progression, I. The basic theorem; II. The Deuring-Heilbronn's phenomenon*, Math. Sb. 15 (1944), 139-178 and 347-368.

[LS]   W. Luo and P. Sarnak, *Quantum variance for Hecke eigenforms*, Ann. Sci. Ecole Norm. Sup. (4) 37 (2004), 769-799.

132   Henryk Iwaniec

[M]   H. L. Montgomery, *The pair correlation of zeros of the zeta function*, Proc.
        Sympos. Pure Math. 24, Amer. Math. Soc. 1973, 181-193.

[MO]   H. L. Montgomery and A. M. Odlyzko, *Gaps between zeros of the zeta function*, Topics in classical number theory (Budapest 1981), Colloq. Math. Soc.
        János Bolyai 34, North-Holland, Amsterdam 1984, 1079-1106.

[MW]   H. L. Montgomery and P. J. Weinberger, *Notes on small class numbers*, Acta
        Arith. 24 (1974), 529-542.

[O]   J. Oesterlé, *Nombres de classes des corps quadratiques imaginaires*, Sém. N.
        Bourbaki (1983-84), exp. 631, Astérisque no. 121-122 (1985), 309-323.

[R-V]   F. Rodríguez Villegas, *Square root formulas for central values of Hecke
        L-series, II,* Duke Math. J. 72 (1993), 431-440.

[SZ]   P. Sarnak and A. Zaharescu, *Some remarks on Landau-Siegel zeros*, Duke
        Math. J. 111 (2002), 495-507.

[S]   C. L. Siegel, *Über die Classenzahl quadratischer Zahlkörper*, Acta Arith. 1
        (1936), 83-86.

[St]   H. M. Stark, *A complete determination of the complex quadratic fields of
        class-number one*, Michigan Math. J. 14 (1967), 1-27.

[Wa]   J.-L. Waldspurger, *Sur les coefficients de Fourier des formes modulaires de
        poids demi-entier*, J. Math. Pures Appl. (9) 60 (1981), 375-484.

Axiomatic Theory of L-Functions: the Selberg Class

Jerzy Kaczorowski

Faculty of Mathematics and Computer Science, Adam Mickiewicz University
ul. Umultowska 87, 61-614 Poznań, Poland
e-mail: kjerzy@amu.edu.pl

L-functions are among the most powerful tools in analytic number theory. It was observed long time ago that most L-functions used in number theory share some common analytic properties. The basic properties are meromorphic continuation, functional equation of Riemann type and Euler product. In the study of known examples certain expectations have appeared. For instance we expect that an analogue of the Riemann hypothesis holds for all sufficiently regular L-functions. A proof of this celebrated hypothesis would have important consequences for number theory. Though very incomplete, our present knowledge of analytic properties of L-functions has already a strong impact on arithmetic.

In order to motivate the study of the Selberg class, we collect in Chapter 1 some examples of concrete L-functions. The main object of these notes are the Selberg classes S and $S^{\#}$, axiomatically defined in Chapter 2. They contain most of the important L-functions used in practice, some of them modulo well-known conjectures. In these lecture notes I have tried to show how far one can go with the explicit description of the structure of the Selberg classes. The ultimate goal of the research in this direction would be a general converse theorem saying that all L-functions from the Selberg class can be obtained as Mellin transforms of automorphic forms associated with arithmetic groups.

We focus on unconditional results. There is a large collection of related open problems and conjectures. We avoid as far as possible a discussion of them.

These lecture notes were prepared for a course delivered at a C.I.M.E. summer school on analytic number theory in Cetraro (Italy), July 2002. I thank the organizers, prof. Alberto Perelli and prof. Carlo Viola, for the invitation. I also thank all the participants in the summer school for their interest in the subject. Special thanks are due to Alberto Perelli and Carlo Viola who read the first version of these notes and made a number of valuable comments.

1 Examples of *L*-functions

1.1 Riemann zeta-function and Dirichlet *L*-functions

For $\sigma > 1$ the Riemann zeta function is defined by the absolutely convergent Dirichlet series

$$\zeta(s) = \sum_{n=1}^{\infty} \frac{1}{n^s}$$

and by analytic continuation elsewhere. The only singularity is a simple pole at $s = 1$ with residue 1. For $\sigma > 1$ we have the following *Euler product formula*

$$\zeta(s) = \prod_{p} \left(1 - \frac{1}{p^s}\right)^{-1}, \tag{1.1}$$

where p runs over the primes. It is well known that it is equivalent to the unique factorization in the ring of integers $\mathbb{Z}$. The Riemann zeta function satisfies the functional equation

$$\pi^{-\frac{s}{2}} \, \Gamma\left(\frac{s}{2}\right) \zeta(s) = \pi^{-\frac{1-s}{2}} \, \Gamma\left(\frac{1-s}{2}\right) \zeta(1-s),$$

where $\Gamma(s)$ is the Euler gamma function.

Using the Mellin inversion formula one shows that the above functional equation is equivalent to the following automorphic property

$$\theta(1/x) = x^{1/2}\theta(x)$$

of the familiar elliptic theta series

$$\theta(x) = \sum_{n\in\mathbb{Z}} e^{-\pi n^2 x} \qquad (x > 0).$$

It is a clear evidence that *L*-functions and automorphic forms should be closely related. This is really the case as we shall see in the next sections.

By (1.1), $\zeta(s)$ has no zeros in the half plane $\sigma > 1$. By the functional equation there are *trivial zeros* at negative even integers. Moreover, there are infinitely many *non-trivial zeros* inside the *critical strip* $0 < \sigma < 1$. The famous Riemann Hypothesis predicts that they all lie on the *critical line* $\sigma = 1/2$.

The definition of $\zeta(s)$ can be generalized as follows. Let $q \geqslant 1$ be an integer and let $\tilde{\chi}$ be a character of the group $(\mathbb{Z}/q\mathbb{Z})^{\times}$ of reduced residue classes (mod q). The induced *Dirichlet character* is a function on the integers defined by the formula

$$\chi(n) = \begin{cases} \tilde{\chi}(n(\operatorname{mod} q)) & \text{if } (n,q) = 1 \\ 0 & \text{otherwise.} \end{cases}$$

If $q' \mid q$ we have the natural group homomorphism

$$(\mathbb{Z}/q\mathbb{Z})^{\times} \longrightarrow (\mathbb{Z}/q'\mathbb{Z})^{\times},$$

and the corresponding injection of the dual groups

$$(\widehat{\mathbb{Z}/q'\mathbb{Z}})^{\times} \longrightarrow (\widehat{\mathbb{Z}/q\mathbb{Z}})^{\times}.$$

A character (mod q), or the induced Dirichlet character, is called *primitive* when it is not in the image of the above map between dual groups for all proper divisors of q. Every Dirichlet character χ (mod q) is induced by a unique primitive Dirichlet character χ^* (mod q^*) for a certain $q^* \mid q$. In this case q^* is called the *conductor* of χ. For every $q \geqslant 1$ there is a unique χ_0 (mod q) with conductor 1, namely

$$\chi_0(n) = \begin{cases} 1 & \text{if } (n, q) = 1 \\ 0 & \text{otherwise,} \end{cases}$$

which is called the *principal character* (mod q).

The Dirichlet L-function associated with a Dirichlet character χ (mod q) is defined for $\sigma > 1$ by the absolutely convergent Dirichlet series

$$L(s, \chi) = \sum_{n=1}^{\infty} \frac{\chi(n)}{n^s}$$

and by analytic continuation elsewhere. The Euler product expansion is

$$L(s, \chi) = \prod_{p} \left(1 - \frac{\chi(p)}{p^s} \right)^{-1}.$$

For the principal character (mod q) we have

$$L(s, \chi_0) = \prod_{p \mid q} \left(1 - \frac{1}{p^s} \right) \zeta(s),$$

whence the analytic properties of $L(s, \chi_0)$ easily follow from the corresponding properties of the Riemann zeta function. For non-principal characters, $L(s, \chi)$ is entire. In the case of a primitive character it satisfies the functional equation

$$\Phi(s, \chi) = \omega_\chi \Phi(1 - s, \overline{\chi}), \tag{1.2}$$

where

$$\Phi(s, \chi) = \left(\frac{q}{\pi} \right)^{s/2} \Gamma\left(\frac{s + a(\chi)}{2} \right) L(s, \chi),$$

$$a(\chi) = \begin{cases} 0 & \text{if } \chi(-1) = 1 \\ 1 & \text{otherwise,} \end{cases}$$

$$\omega_\chi = \frac{\tau(\chi)}{i^{a(\chi)}\sqrt{q}}$$

and $\tau(\chi)$ denotes the corresponding Gaussian sum. We have $|\omega_\chi| = 1$.

As in the case of the Riemann zeta function, $L(s, \chi)$ does not vanish in the half plane $\sigma > 1$. For $q > 1$ and primitive χ, the trivial zeros are at the points

$$-2k - a(\chi), \qquad k \geqslant 0.$$

The non-trivial zeros are in the critical strip, and the Generalized Riemann Hypothesis predicts that they all lie on the critical line.

Some linear combinations of Dirichlet L-functions satisfy functional equations of a similar type. The best known example is the Davenport-Heilbronn zeta function:

$$L(s) = \overline{\lambda} L(s, \chi_1) + \lambda L(s, \overline{\chi}_1),$$

where χ_1 is the complex character (mod 5) such that $\chi_1(2) = i$, and

$$\lambda = \frac{1}{2}\left(1 + i\frac{\sqrt{10 - 2\sqrt{5}} - 2}{\sqrt{5} - 1}\right).$$

Its functional equation is

$$\left(\frac{\pi}{5}\right)^{\frac{s}{2}} \Gamma\left(\frac{s+1}{2}\right) L(s) = \left(\frac{\pi}{5}\right)^{\frac{1-s}{2}} \Gamma\left(\frac{2-s}{2}\right) L(1-s).$$

The Dirichlet coefficients of $L(s)$ are not multiplicative and hence we have no Euler product in this case. One can show that $L(s)$ badly violates the Riemann Hypothesis: there are infinitely many zeros in the half-plane $\sigma > 1$.

Further reading: [10], [15], [36], [43].

1.2 Hecke L-functions

Let K be a finite extension of the field of rational numbers $\mathbb{Q}$ of degree $n = r_1 + 2r_2$ and discriminant D_K. Let J_K denote the group of ideles of K with its natural topology. We identify K^* with the diagonal of J_K. Let S be a finite set of primes of K containing all infinite (Archimedean) primes. Moreover, let ψ be an idele class group character, i.e. a continuous homomorphism

$$\psi : J_K \longrightarrow \mathbb{C}^*$$

which is trivial on K^* and satisfies $\psi(x) = 1$ if $x_v = 1$ for $v \in S$ and $|x_v| = 1$ for $v \notin S$.

Every idele class character induces a character of the group I_K^S of the fractional ideals of K generated by all primes $v \notin S$. This is done as follows. Let $I \in I_K^S$,

$$I = \prod_{v \notin S} \mathfrak{p}_v^{\alpha_v},$$

where $\mathfrak{p}_v$ is the prime ideal associated with v and the α_v are integers. We associate the idele x^I as follows:

$$x_v^I = \begin{cases} 1 & \text{if } v \in S \\ \pi_v^{\alpha_v} & \text{if } v \notin S, \end{cases}$$

where for $v \notin S$, π_v denotes the local uniformizing element for v. Then we put

$$\chi(I) = \psi(x^I).$$

Note that x^I is not uniquely determined by I due to the freedom in choosing the π_v's, but since ψ is an idele class group character, $\psi(x^I)$ is the same for all possible choices. We extend the definition of χ to all ideals of K putting $\chi(I) = 0$ if $I \notin I_K^S$. We call every χ defined in this way a *Hecke character*.

Let S and S' be two finite sets of primes containing all infinite primes. Two Hecke characters χ and χ' defined on I_K^S and $I_K^{S'}$ are *equivalent* when $\chi(I) = \chi'(I)$ for every $I \in I_K^S \cap I_K^{S'}$. For every χ there exists a unique χ^* equivalent to χ which is defined for the smallest possible S. If $\chi = \chi^*$ then χ is called *primitive*.

Hecke characters admit an explicit description.

Let $\mathfrak{f}$ be a non-zero integral ideal of K and let $H_{\mathfrak{f}}^*$ be the corresponding ideal class group (mod $\mathfrak{f}$) in narrow sense. Every character χ of $H_{\mathfrak{f}}^*$ induces a Hecke character as follows:

$$\chi(I) = \begin{cases} \chi([I]_{\mathfrak{f}}) & \text{if } (I, \mathfrak{f}) = 1 \\ 0 & \text{otherwise,} \end{cases}$$

where $[I]_{\mathfrak{f}}$ denotes the class of I in $H_{\mathfrak{f}}^*$. Every such character has finite order, and conversely every Hecke character of finite order can be obtained in this way.

If $\mathfrak{f}' \mid \mathfrak{f}$ then we have the natural injection $\widehat{H_{\mathfrak{f}'}^*} \to \widehat{H_{\mathfrak{f}}^*}$. Primitive Hecke characters (mod $\mathfrak{f}$) are those which are not obtained in this way by proper divisors of $\mathfrak{f}$. Every Hecke character of finite order χ (mod $\mathfrak{f}$) is induced by the unique primitive character (mod $\mathfrak{f}^*$) for certain $\mathfrak{f}^* \mid \mathfrak{f}$. We call $\mathfrak{f}^*$ the *conductor* of χ.

Hecke characters of infinite order are more complicated. Let $\mathfrak{f}$ be an integral ideal of K and $U_{\mathfrak{f}}^+(K)$ and $\mu_{\mathfrak{f}}(K)$ denote the group of totally positive units $\equiv 1$ (mod $\mathfrak{f}$) and the group of roots of unity $\equiv 1$ (mod $\mathfrak{f}$) respectively. Let $\varepsilon_1, \ldots, \varepsilon_r$ denote a system of fundamental units of $U_{\mathfrak{f}}^+(K)$ and let ζ be a generator of $\mu_{\mathfrak{f}}(K)$. For each Archimedean v we choose an integer n_v ($n_v \in \{0, 1\}$ if v is real) and a real number t_v such that

$$\prod_v \zeta_v^{n_v} = 1$$

and the sum

$$\sum_v t_v \log |(\varepsilon_j)_v| + \sum_{v \text{ complex}} n_v \arg (\varepsilon_j)_v$$

is an integral multiple of 2π for every $j = 1, \ldots, r$. Then for every non-zero $x \in K$ which is totally positive and $\equiv 1 \pmod{\mathfrak{f}}$ we put

$$f(x) = \prod_v |x_v|^{it_v} \prod_{v \text{ complex}} (x_v/|x_v|)^{n_v}. \tag{1.3}$$

Every such f is called *Grössencharacter*. Let the ideal class group $H_{\mathfrak{f}}^*$ decompose into cyclic factors of orders $h_1, \ldots, h_N$ and let $J_1, \ldots, J_N$ denote ideals whose classes are the corresponding generators. Of course every $J_j^{h_j}$, $j = 1, \ldots, N$, is principal:

$$J_j^{h_j} = (x_j), \qquad x_j \equiv 1 \pmod{\mathfrak{f}}, \qquad x_j \text{ totally positive.}$$

For every ideal I of the form

$$I = x J_1^{b_1} \ldots J_N^{b_N},$$

with $x \equiv 1 \pmod{\mathfrak{f}}$, totally positive, and $0 \leqslant b_j < h_j$ for $j = 1, \ldots, N$, we put

$$\chi(I) = f(x) w_1^{b_1} \ldots w_N^{b_N} \psi(I), \tag{1.4}$$

where ψ is a character of $H_{\mathfrak{f}}^*$ and w_j, $j = 1, \ldots, N$, are fixed h_j-th roots of $f(x_j)$. We extend the definition of χ to all ideals putting $\chi(I) = 0$ if I is not prime to $\mathfrak{f}$. Every such function is a Hecke character and every Hecke character can be obtained in this way.

When

$$\sum_v t_v = 0$$

we call χ *normalized*. Moreover, χ is primitive when the corresponding finite order character ψ is primitive.

For a normalized Hecke character χ we define the corresponding Hecke L-function for $\sigma > 1$ by the absolutely convergent Dirichlet series

$$L_K(s, \chi) = \sum_{I \neq 0} \frac{\chi(I)}{N(I)^s},$$

where $N(I)$ denotes the norm of I, and by analytic continuation elsewhere. Summation is over all non-zero integral ideals of K. When K is fixed and no confusion can occur, we simply write $L(s, \chi)$ instead of $L_K(s, \chi)$.

Existence of analytic continuation was established by E. Hecke who also proved that $L(s, \chi)$ with primitive χ satisfies the following functional equation

$$\Phi(s, \chi) = \omega(\chi) \Phi(1 - s, \overline{\chi}), \tag{1.5}$$

where

$$\Phi(s,\chi) = \left(\frac{N(\mathfrak{f})\,|D_K|}{4^{r_2}\pi^n}\right)^{s/2} \times$$

$$\prod_{v \text{ real}} \Gamma\left(\frac{s + n_v + it_v}{2}\right) \prod_{v \text{ complex}} \Gamma\left(s + \frac{|n_v|}{2} + it_v\right) L(s,\chi), \quad (1.6)$$

and $\omega(\chi)$ is a complex number with $|\omega(\chi)| = 1$.

The *root number* $\omega(\chi)$ can be determined explicitly. For every prime v let $\chi_v = \chi_{|K_v^*}$. Then $\chi = \prod_v \chi_v$ and

$$\omega(\chi) = \prod_v \omega(\chi_v). \qquad (1.7)$$

Moreover, we have

$$\omega(\chi_v) = \begin{cases} 1 & \text{if } K_v = \mathbb{C}, \\ 1 & \text{if } \chi_v = 1 \text{ and } K_v = \mathbb{R}, \\ i & \text{if } \chi_v \neq 1 \text{ and } K_v = \mathbb{R}, \\ N(\mathfrak{f}(\chi_v))^{-1/2}\tau(\overline{\chi_v}) & \text{for non-Archimedean } v. \end{cases}$$

Here $\mathfrak{f}(\chi_v)$ denotes the local conductor, N is the absolute norm and τ is the generalized Gaussian sum:

$$\tau(\overline{\chi_v}) = \sum_{x \in \mathcal{O}_v^* \,(\mathrm{mod}^*\,\mathfrak{f}(\chi_v))} \overline{\chi_v}(d^{-1}x)\psi_v(d^{-1}x),$$

where d is a generator of $\mathfrak{f}(\chi_v)D_v$ (D_v is the absolute different of K_v), ψ_v is the canonical character of K_v^+ and the summation is over a set of representatives of the cosets $1 + \mathfrak{f}(\chi_v)$ in $\mathcal{O}_v^*$ ($\mathcal{O}_v$ is the ring of integers of K_v).

There is one important special case in which χ is trivial. Then the corresponding Hecke L-function is, for $\sigma > 1$,

$$\zeta_K(s) = \sum_{I \neq 0} N(I)^{-s}$$

and is called the *Dedekind zeta function* of K. The only singularity is a simple pole at $s = 1$ with residue

$$\kappa_K = h\,\frac{2^{r_1}(2\pi)^{r_2}R_K}{\sqrt{|D_K|}\,w_K},$$

where h, R_K and w_K denote the class number, the regulator and the number of roots of unity of K respectively. Hecke L-functions with non-trivial characters are entire.

Since the characters are multiplicative functions over ideals, we have the Euler product

$$L(s, \chi) = \prod_{\mathfrak{p}} \left(1 - \frac{\chi(\mathfrak{p})}{N(\mathfrak{p})^s} \right)^{-1}, \tag{1.8}$$

where $\mathfrak{p}$ runs over all non-zero prime ideals of K. Since every prime ideal lies over a rational prime and each rational prime splits into a product of at most n prime ideals, (1.8) can be written in the form

$$L(s, \chi) = \prod_p L_p(s, \chi),$$

where p runs over rational primes and the inverse of every local factor $L_p(s, \chi)$ is a polynomial of degree $\leqslant n$ at p^{-s}:

$$L_p(s, \chi) = \prod_{j=1}^n \left(1 - \frac{\alpha_{j,p}}{p^s} \right)^{-1}$$

and $|\alpha_{j,p}| \leqslant 1$.

In the case where $K = \mathbb{Q}$ is the field of rationals, every Hecke character is of the form

$$\chi(n) n^{it},$$

where χ is a Dirichlet character and t is a real number. Such a Hecke character is normalized when $t = 0$ and then the corresponding Hecke L-function is exactly the Riemann zeta function or the Dirichlet L-function.

Further reading: [14], [33], [42].

1.3 Artin L-functions

Let K/k be a normal extension of algebraic number fields with Galois group G and rings of integers $\mathcal{O}_K$ and $\mathcal{O}_k$ respectively. Denote by ρ a finite dimensional representation of G in a vector space V. Moreover, let χ denote its character.

Given a prime $\mathfrak{p}$ of k we choose a prime $\mathfrak{P}$ of K lying above $\mathfrak{p}$. Let $D_{\mathfrak{P}} := \{ \sigma \in G : \mathfrak{P}^\sigma = \mathfrak{P} \}$ and $I_{\mathfrak{P}} := \{ \sigma \in G : \sigma(a) \equiv a \pmod{\mathfrak{P}} \text{ for every } a \in \mathcal{O}_K \}$ denote the decomposition and inertia group respectively. The quotient group $D_{\mathfrak{P}}/I_{\mathfrak{P}}$ is canonically isomorphic to the Galois group of $\mathcal{O}_k/\mathfrak{p} \subset \mathcal{O}_K/\mathfrak{P}$. Let $\sigma_{\mathfrak{P}} = \left[\frac{K/k}{\mathfrak{P}} \right]$ denote the corresponding Frobenius substitution. We write

$$V^{I_{\mathfrak{P}}} := \{ \mathrm{v} \in V : \rho(\sigma)(\mathrm{v}) = \mathrm{v} \text{ for every } \sigma \in I_{\mathfrak{P}} \}.$$

We define the local Artin L-function corresponding to a finite prime $\mathfrak{p}$ of k by the formula

$$L_{\mathfrak{p}}(s, K/k, \rho) = \left(\det \left(I - N(\mathfrak{p})^{-s} \rho(\sigma_{\mathfrak{P}}) \right) \right)^{-1}, \tag{1.9}$$

where I denotes the unit matrix of dimension $\dim V^{I_{\mathfrak{P}}}$ and s denotes a complex number with a positive real part. One checks without difficulty that the right-hand side of (1.9) does not depend on the particular choice of $\mathfrak{P}$ above $\mathfrak{p}$ and that it is the same for all equivalent representations. Therefore $L_{\mathfrak{p}}(s, K/k, \rho)$ depends only on χ and we can write $L_{\mathfrak{p}}(s, K/k, \chi)$ instead of $L_{\mathfrak{p}}(s, K/k, \rho)$.

Let us fix a rational prime p and consider the product $L_p(s, K/k, \chi)$ of all local Artin L-functions taken over all finite primes $\mathfrak{p}$ of k lying above p:

$$L_p(s, K/k, \chi) = \prod_{\mathfrak{p}|p} L_{\mathfrak{p}}(s, K/k, \chi).$$

Suppose for simplicity that p is unramified in $K/\mathbb{Q}$. Then $V^{I_{\mathfrak{p}}} = V$ for every $\mathfrak{p}|p$. Moreover, we can assume without loss of generality that $\rho(\sigma_{\mathfrak{P}})$ is represented by a diagonal matrix

$$\begin{pmatrix} \varepsilon_1 & & 0 \\ & \ddots & \\ 0 & & \varepsilon_n \end{pmatrix}$$

($n = \dim V$). Since G is a finite group, the ε_j's are roots of unity. Therefore

$$L_{\mathfrak{p}}(s, K/k, \chi) = \prod_{j=1}^{n} \left(1 - \varepsilon_j N(\mathfrak{p})^{-s}\right)^{-1}.$$

If $\mathfrak{p}_1, \ldots, \mathfrak{p}_t$ are all primes of k lying above p, then $N(\mathfrak{p}_j) = p^{f_j}$ for every $j = 1, \ldots, t$ and $\sum_{j=1}^{t} f_j = [k : \mathbb{Q}]$. Hence

$$L_p(s, K/k, \chi) = \prod_{j=1}^{n[k:\mathbb{Q}]} \left(1 - \zeta_{j,p} p^{-s}\right)^{-1} \tag{1.10}$$

for certain roots of unity $\zeta_{j,p}$. Thus $L_p(s, K/k, \chi)$ is the inverse of a polynomial of degree $[k : \mathbb{Q}] \dim V$ at p^{-s}. The roots of the polynomial in question are roots of unity. For the finite number of ramifying primes p we have a similar statement, but in these cases the degrees of the involved polynomials are less than $[k : \mathbb{Q}] \dim V$.

The global *Artin L-function* $L(s, K/k, \chi)$ is defined as the product of all local factors:

$$L(s, K/k, \chi) = \prod_{p} L_p(s, K/k, \chi). \tag{1.11}$$

The product converges for $\Re s > 1$ and hence $L(s, K/k, \chi)$ is holomorphic in this half-plane.

Expanding the local factors in (1.10), one can write $L(s, K/k, \chi)$ for $\Re s > 1$ as an absolutely convergent Dirichlet series

$$\sum_{n=1}^{\infty} \frac{a(n)}{n^s},$$

say. The absolute values of the coefficients are bounded by an appropriate divisor function and therefore the following *Ramanujan condition* holds:

$$a(n) \ll n^{\varepsilon}$$

for every positive ε.

Theorem 1.3.1 (Artin) *We have:*

1. *For two characters χ_1 and χ_2 of G we have $L(s, K/k, \chi_1 + \chi_2) = L(s, K/k, \chi_1)\, L(s, K/k, \chi_2)$.*

2. *If H is a subgroup of G and E denotes the corresponding field, then for every character χ of H*

$$L(s, K/E, \chi) = L(s, K/k, \mathrm{Ind}_H^G(\chi)),$$

   where $\mathrm{Ind}_H^G(\chi)$ denotes the induced character of G.

3. *If H is a normal subgroup of G then every character χ of the quotient group G/H defines in a canonical way a character χ' of G, and*

$$L(s, E/k, \chi) = L(s, K/k, \chi').$$

4. *(Artin's reciprocity law) If K/k is abelian then for every character χ of G there exists an ideal $\mathfrak{f}$ of $\mathcal{O}_k$ and a character χ^* of the ideal class group $H_{\mathfrak{f}}^*(k)$ such that*

$$L(s, K/k, \chi) = L_k(s, \chi^*),$$

   where $L_k(s, \chi^*)$ denotes the Hecke L-function of k associated with χ^* (see Section 1.2).

The first property reduces the study of Artin L-functions to the case of irreducible representations. The last property provides the analytic continuation of all abelian Artin L-functions. Using *3* we can define the Artin L-functions for every (virtual) character of $\mathrm{Gal}(\overline{\mathbb{Q}}/\mathbb{Q})$.

Let us take $H = \{1\}$ in *2*. Then the induced representation is just the regular representation of G, and the induced character is

$$\sum_{\chi} (\dim \chi) \chi,$$

where the sum is over all irreducible characters of G. Since $L(s, K/K, 1)$ is the Dedekind zeta function of K, as a consequence we obtain

$$\zeta_K(s) = \prod_{\chi} L(s, K/k, \chi)^{\dim \chi}. \tag{1.12}$$

Artin's Conjecture *Every $L(s, K/k, \chi)$, where χ is the character of an irreducible representation, admits meromorphic continuation to the whole complex plane. It is entire if $\chi \neq 1$ and has a simple pole at $s = 1$ otherwise.*

The most successful approach to this conjecture uses Theorem 1.3.1 and a theorem of Brauer (see for instance [39], Theorem 23) stating that every character of a finite group is a linear combination with integer coefficients of characters induced by characters of degree 1. Hence by Theorem 1.3.1, *2*, we can write

$$L(s, K/k, \chi) = \prod_{j=1}^{J} L^{n_j}(s, K/E_j, \chi_j) \tag{1.13}$$

for certain intermediate fields $k \subset E_j \subset K$, for degree one characters χ_j of groups $G_j = \text{Gal}(K/E_j)$, and certain integers n_j. Every character of degree one factors through G_j^{ab}, the quotient of G_j by its commutator subgroup. Hence using Theorem 1.3.1, *3*, we see that the corresponding factors on the right-hand side of (1.13) coincide with the Artin L-functions of certain abelian extensions F_j/E_j with $F_j \subset K$ and hence, by the Artin reciprocity law, they are Hecke L-functions.

We see therefore that every Artin L-function can be written as a quotient of products of Hecke L-functions associated with finite order Hecke characters of certain intermediate fields $k \subset E_j \subset K$. In particular it admits meromorphic continuation to the whole complex plane and satisfies a functional equation with multiple gamma factors.

Let us consider more carefully the problem of functional equation.

Let v be a real infinite prime of k and let w be an infinite prime of K lying above v. Let σ_w denote the generator of the inertia group $G(w) = \{\sigma \in G : \sigma w = w\}$. Note that $G(w)$ is cyclic of order at most 2 and hence σ_w exists. The matrix $\rho(\sigma_w)$ has at most two eigenvalues $+1$ or -1. Accordingly V splits into the direct sum of two subspaces $V = V_v^+ \oplus V_v^-$.

For complex s we write

$$g(s) = \pi^{-\frac{s}{2}} \Gamma\left(\frac{s}{2}\right).$$

Then for every infinite prime v of k let

$$\gamma_v(s) = \begin{cases} g(s)^{\dim V} g(s+1)^{\dim V} & \text{if } v \text{ is complex,} \\ g(s)^{\dim V_v^+} g(s+1)^{\dim V_v^-} & \text{if } v \text{ is real.} \end{cases}$$

We define the *gamma factor* of χ as follows:

$$\gamma_\chi = \prod_v \gamma_v(s),$$

where the product is taken over all infinite primes of k.

In order to define the gamma factor of $L(s, K/k, \chi)$ and write the functional equation, we have to introduce the Artin conductor of χ. We proceed locally. Let $\mathfrak{p}$ be a prime of k and let $\mathfrak{P}$ be a prime of K lying above $\mathfrak{p}$. We denote by G_i $(i \geqslant 0)$ the corresponding ramification groups. Write

$$n(\chi, \mathfrak{p}) = \sum_{i=0}^{\infty} \frac{|G_i|}{|G_0|} \operatorname{codim} V^{G_i}.$$

Artin proved that this is an integer. We have $n(\chi, \mathfrak{p}) = 0$ for unramified $\mathfrak{p}$. Hence the following product

$$f(\chi, K/k) = \prod_{\mathfrak{p}} \mathfrak{p}^{n(\chi, \mathfrak{p})}$$

is well defined and represents an ideal of k, called the *Artin conductor*.

Theorem 1.3.2 (Artin) *The complete Artin L-function*

$$\Lambda(s, K/k, \chi) = A(\chi)^{s/2} \gamma_\chi(s) L(s, K/k, \chi),$$

where

$$A(\chi) = |D_k|^{\dim V} N_{k/\mathbb{Q}}(f(\chi, K/k))$$

and D_k denotes the absolute discriminant of k, satisfies the following functional equation

$$\Lambda(1 - s, K/k, \chi) = W(\chi) \Lambda(s, K/k, \overline{\chi}),$$

for some constant $W(\chi)$ of absolute value 1 (the Artin root number).

As we have already seen, every Artin L-function can be expressed as a product of Hecke L-functions. If

$$L(s, K/k, \chi) = \prod_{j=1}^{J} L_{E_j}^{n_j}(s, \chi_j),$$

where the E_j's are intermediate fields $(k \subset E_j \subset K)$, the χ_j's are Hecke characters of finite order and the n_j's are integers, then

$$W(\chi) = \prod_{j=1}^{J} \omega(\chi_j)^{n_j}$$

(see (1.7)). Hence the Artin root number is expressed in terms of root numbers of Hecke L-functions and therefore in terms of generalized Gaussian sums. It follows in particular that $W(\chi)$ is always an algebraic number.

One can wonder about exact relations between Artin and Hecke L-functions. We know that these theories overlap. For instance every Hecke character of finite order generates an Artin L-function. But not all Artin L-functions arise in this way. Also there are Hecke L-functions which are not Artin L-functions. There exists however a unifying theory coming back to A. Weil, who attached to every algebraic number field K a topological group W_K, nowadays called the (*absolute*) *Weil group*, and proved that every representation of W_K induces an L-series having meromorphic continuation and functional equation. The construction is in fact a generalization of Artin's theory. Moreover, we have a homomorphism $W_K \longrightarrow \mathrm{Gal}(\overline{K}/K)$ and hence every Galois representation induces a representation of the Weil group. Consequently, every Artin L-function can be obtained in this way. On the other hand, one obtains all Hecke L-functions as well (this follows from the class field theory).

Further reading: [14], [28].

1.4 GL$_2$ L-functions

1.4.1 Definitions and basic properties

Let $H = \{z \in \mathbb{C} : \Im z > 0\}$ denote the upper half-plane. The group

$$\mathrm{SL}_2(\mathbb{R}) = \left\{ \begin{pmatrix} a & b \\ c & d \end{pmatrix} : a, b, c, d \in \mathbb{R}, \ ad - bc = 1 \right\}$$

acts on H by the linear transformations

$$\gamma z = \frac{az + b}{cz + d}, \quad \text{where } \gamma = \begin{pmatrix} a & b \\ c & d \end{pmatrix}.$$

It also acts on the space of holomorphic functions on H as follows:

$$f|_k \gamma(z) := (cz + d)^{-k} f(\gamma z) \qquad (\gamma \in \mathrm{SL}_2(\mathbb{R})),$$

for every fixed integer k.

Let N be a positive integer and denote by $\Gamma_0(N)$ the *Hecke congruence subgroup* of level N:

$$\Gamma_0(N) = \left\{ \gamma = \begin{pmatrix} a & b \\ c & d \end{pmatrix} \in \mathrm{SL}_2(\mathbb{Z}) : c \equiv 0 \,(\mathrm{mod}\,N) \right\}.$$

Let χ be a Dirichlet character (mod N). A *modular form of type* (k, χ) and *level* N is a function on H such that

$$f|_k \gamma = \chi(d) f$$

for every $\gamma = \begin{pmatrix} a & b \\ c & d \end{pmatrix} \in \Gamma_0(N)$, and which is holomorphic on H and at the cusps of $\Gamma_0(N)$. Holomorphy at cusps needs explanation. Since $\begin{pmatrix} 1 & 1 \\ 0 & 1 \end{pmatrix} \in \Gamma_0(N)$, f is 1-periodic. Hence we have the Fourier series expansion

$$f(z) = \sum_{n=-\infty}^{\infty} a(n)e(nz) \qquad (e(x) := e^{2\pi i x}). \tag{1.14}$$

We say that f is holomorphic at $i\infty$ (cusp at infinity) if the coefficients in (1.14) vanish for $n < 0$. This definition is a natural one if we take into account that $e(z)$ is the local uniformizer at $i\infty$ and (1.14) is the power series expansion in $e(z)$. If A is another cusp then we first transform f by the appropriate scaling matrix γ_A which sends $i\infty$ to A and then apply the former definition. Hence f is holomorphic at A if

$$f|_k\gamma_A(z) = \sum_{n=0}^{\infty} a(n, A)e(nz).$$

If in addition $a(0, A) = 0$ for all cusps A, then f is called a *cusp form*. If a modular form of type (k, χ) exists then the *weight* k and the Dirichlet character χ satisfy the following consistency condition

$$\chi(-1) = (-1)^k. \tag{1.15}$$

Let $\mathcal{S}(N, k, \chi)$ denote the complex vector space of cusp forms of type (k, χ) and level N. Let D be a fundamental domain for $\Gamma_0(N)$. We make $\mathcal{S}(N, k, \chi)$ to be a Hilbert space with the following *Petersson inner product*

$$\langle f, g \rangle = \iint_D f(z)\overline{g(z)}\, y^k \frac{\mathrm{d}x\,\mathrm{d}y}{y^2} \qquad (z = x + iy).$$

Let $f(z) = \sum_{n=1}^{\infty} a(n)e(nz)$ be a cusp form of weight k. It is easy to prove that its Fourier coefficients satisfy the following inequality:

$$a(n) = O(n^{k/2}). \tag{1.16}$$

We define the L-function of f as an absolutely convergent Dirichlet series

$$L(s, f) = \sum_{n=1}^{\infty} \frac{a(n)}{n^s} \qquad (\sigma = \Re s > (k+2)/2).$$

It is holomorphic for $\sigma > (k+2)/2$, it extends to an entire function on $\mathbb{C}$ and satisfies the following functional equation

$$\Lambda(s, f) = i^k \Lambda(k - s, f|_k\omega_N), \tag{1.17}$$

where

$$\Lambda(s, f) = \left(\frac{2\pi}{\sqrt{N}}\right)^{-s} \Gamma(s)\, L(s, f)$$

and

$$\omega_N = \begin{pmatrix} 0 & -1 \\ N & 0 \end{pmatrix}.$$

We remark here that if $f \in \mathcal{S}(N, k, \chi)$, then $f|_k \omega_N \in \mathcal{S}(N, k, \overline{\chi})$.

If $f(z) = \sum_{n=0}^{\infty} a(n)e(nz)$ is a modular form of type (k, χ) and level N and ψ is a primitive Dirichlet character $(\mathrm{mod}\, r)$ then the series

$$f^{\psi}(z) := \sum_{n=0}^{\infty} \psi(n)a(n)e(nz)$$

defines a modular form of type $(k, \chi\psi^2)$ and level M, where M denotes the least common multiple of N, $N^* r$ and r^2, $N^*|N$ being the conductor of χ. Moreover, if f is a cusp form, so is f^{ψ}. Consequently, if $f \in \mathcal{S}(N, k, \chi)$ then the *twisted L-function* defined for $\sigma > (k + 2)/2$ by

$$L^{\psi}(s, f) = L(s, f^{\psi}) = \sum_{n=1}^{\infty} \frac{\psi(n)a(n)}{n^s}$$

extends to an entire function on $\mathbb{C}$ and satisfies a functional equation of the usual type. More precisely if (for simplicity) $(N, r) = 1$ then

$$\Lambda(s, f, \psi) = i^k w(\psi) \Lambda(k - s, f|_k \omega_N, \overline{\psi}), \tag{1.18}$$

where

$$\Lambda(s, f, \psi) = \left(\frac{r\sqrt{N}}{2\pi}\right)^s \Gamma(s) L^{\psi}(s, f)$$

and

$$w(\psi) = \chi(r)\psi(N)\tau^2(\psi)r^{-1}, \tag{1.19}$$

$\tau(\psi)$ denoting the corresponding Gaussian sum.

1.4.2 Hecke operators and newforms

Let p be a prime number. We define the Hecke operator T_p acting on $\mathcal{S}(N, k, \chi)$ as follows:

$$T_p(f)(z) = p^{k-1}f(pz) + \chi(p)\frac{1}{p}\sum_{b\,(\mathrm{mod}\,p)} f\left(\frac{z+b}{p}\right).$$

Theorem 1.4.1 (Hecke) *Let $f(z) = \sum_{n=1}^{\infty} a(n)e(nz) \in \mathcal{S}(N, k, \chi)$ be normalized ($a(1) = 1$). Then the following two statements are equivalent:*

1. *f is a common eigenfunction of all Hecke operators.*

2. *The associated L-function $L(s, f)$ has an Euler product of the form*

$$L(s, f) = \prod_{p \nmid N} \left(1 - a(p)p^{-s} + \chi(p)p^{k-1-2s}\right)^{-1} \prod_{p | N} \left(1 - \frac{a(p)}{p^s}\right)^{-1}.$$

Suppose that N' and N are two different positive integers, $N'|N$, and let $f \in \mathcal{S}(N', k, \chi)$ for a certain Dirichlet character $\chi \pmod{N'}$. Then for every positive integer d satisfying $dN'|N$ the function

$$f^*(z) := f(dz)$$

belongs to $\mathcal{S}(N, k, \chi)$. Every such f^* is called an *oldform*. They span a subspace $\mathcal{S}^-(N, k, \chi)$ of $\mathcal{S}(N, k, \chi)$. Let $\mathcal{S}^+(N, k, \chi)$ be its orthogonal complement under the Petersson inner product. This space is spanned by non-zero cusp forms which are eigenfunctions of all the Hecke operators, the so-called *newforms*. If $f \in \mathcal{S}^+(N, k, \chi)$ is a newform then $a(1) \neq 0$ and

$$f|_k \omega_N(z) = \epsilon \overline{f(-\overline{z})}, \tag{1.20}$$

where ϵ is a constant of absolute value 1. Then $f|_k \omega_N$ is a newform from $\mathcal{S}(N, k, \overline{\chi})$ (the bar denotes as usual complex conjugation).

For newforms we have the following celebrated theorem of P. Deligne improving (1.16).

Theorem 1.4.2 (Deligne) *Let $f(z) = \sum_{n=1}^{\infty} a(n)e(nz)$, $a(1) = 1$, be a newform of type (k, χ) and level N. Then for each prime p not dividing N we have*

$$1 - a(p)p^{-s} + \chi(p)p^{k-1-2s} = (1 - \alpha(p)p^{-s})(1 - \beta(p)p^{-s}),$$

where $|\alpha(p)| = |\beta(p)| = p^{\frac{k-1}{2}}$. In particular

$$a(n) \ll n^{\frac{k-1}{2}+\varepsilon}$$

for every $\varepsilon > 0$, and the Dirichlet series of $L(s, f)$ converges absolutely for $\sigma > (k+1)/2$.

1.4.3 Converse theorems

Let us consider the case of level 1 and type (k, χ). Of course χ is identically 1, and by the consistency condition (1.15) k has to be even.

Theorem 1.4.3 (Hecke) *Let k be an even positive integer and let*

$$f(z) = \sum_{n=0}^{\infty} a(n)e(nz)$$

with $a(n) \ll n^c$ for a constant c. Then the following two statements are equivalent.

1. f is a modular form of weight k for the full modular group $\Gamma_0(1)$.

2. The function

$$\Lambda_f(s) = (2\pi)^{-s} \Gamma(s) \sum_{n=1}^{\infty} \frac{a(n)}{n^s}$$

has meromorphic continuation to the whole complex plane,

$$\Lambda_f(s) + \frac{a(0)}{s} + \frac{i^k a(0)}{k-s}$$

is entire and bounded on vertical strips, and $\Lambda_f(s)$ satisfies the following functional equation:

$$\Lambda_f(s) = i^k \Lambda_f(k-s).$$

This is so since $\Gamma_0(1)$ is generated by $\begin{pmatrix} 1 & 1 \\ 0 & 1 \end{pmatrix}$ and $\begin{pmatrix} 0 & 1 \\ -1 & 0 \end{pmatrix}$. A similar theorem is also true for some other small levels, but in general there are more generators and one functional equation does not suffice.

Theorem 1.4.4 (Weil) *Let k and N be two positive integers and let χ be a Dirichlet character (mod N). Suppose we are given a Fourier series $f(z) = \sum_{n=0}^{\infty} a(n)e(nz)$ with coefficients satisfying $a(n) = O(n^A)$ for some positive A. Suppose moreover that the corresponding Dirichlet series*

$$L(s, f) = \sum_{n=1}^{\infty} \frac{a(n)}{n^s}$$

has the following properties.

1. *$L(s, f)$ can be analitically continued to the whole complex plane and satisfies the following functional equation:*

$$\left(\frac{\sqrt{N}}{2\pi} \right)^s \Gamma(s) L(s, f) = w_0 \left(\frac{\sqrt{N}}{2\pi} \right)^{k-s} \Gamma(k-s) \overline{L}(k-s, f)$$

with $w_0 \in \mathbb{C}$, $|w_0| = 1$. Moreover, the function

$$\left(\frac{\sqrt{N}}{2\pi} \right)^s \Gamma(s) L(s, f) + \frac{a(0)}{s} + w_0 \frac{\overline{a(0)}}{k-s}$$

is entire and bounded on every vertical strip.

2. *For almost all primitive Dirichlet characters ψ of prime conductor $r \nmid N$, the twist*

$$L^\psi(s, f) = \sum_{n=1}^{\infty} a(n)\psi(n)n^{-s}$$

extends to an entire function and satisfies the functional equation (see (1.18))

$$\left(\frac{r\sqrt{N}}{2\pi} \right)^s \Gamma(s) L^\psi(s, f) = w_0\, w(\psi) \left(\frac{r\sqrt{N}}{2\pi} \right)^{k-s} \Gamma(k-s) \overline{L^\psi}(k-s, f)$$

where $w(\psi)$ is given by (1.19).

Then f is a modular form of type (k, χ) and level N. Moreover, if $L(s, f)$ converges absolutely for $\sigma > k - \delta$ for some positive δ, then f is a cusp form.

1.4.4 Normalization

Later on we shall need the following normalization of the L-function attached to a newform f. For $f(z) = \sum_{n=1}^{\infty} a(n)e(nz) \in S(N, k, \chi)$ let us denote by $\widetilde{L}(s, f)$ its normalized L-function defined as follows:

$$\widetilde{L}(s, f) = \frac{1}{a(1)} L\left(s + \frac{k-1}{2}, f\right). \tag{1.21}$$

For $\sigma > 1$ we have

$$\widetilde{L}(s, f) = \sum_{n=1}^{\infty} \frac{\widetilde{a}(n)}{n^s}$$

and the series converges absolutely since by Deligne's theorem the following Ramanujan condition holds:

$$\widetilde{a}(n) \ll n^\varepsilon$$

for every $\varepsilon > 0$. Of course $\widetilde{L}(s, f)$ extends to an entire function and satisfies the following functional equation:

$$\widetilde{\Lambda}(s, f) = \omega \, \overline{\widetilde{\Lambda}(1 - \overline{s}, f)},$$

with $\omega = i^k \epsilon$, where ϵ is as in (1.20), and where

$$\widetilde{\Lambda}(s, f) = \left(\frac{2\pi}{\sqrt{N}}\right)^{-s} \Gamma\left(s + \frac{k-1}{2}\right) \widetilde{L}(s, f).$$

The Euler product of $\widetilde{L}(s, f)$ has the following form:

$$\widetilde{L}(s, f) = \prod_{p \nmid N} \left(1 - \frac{\alpha(p)}{p^s}\right)^{-1} \left(1 - \frac{\beta(p)}{p^s}\right)^{-1} \prod_{p \mid N} \left(1 - \frac{a(p)}{p^s}\right)^{-1},$$

with $|\alpha(p)| = |\beta(p)| = 1$. Using a terminology which we shall introduce later, we can say that $\widetilde{L}(s, f)$ has a polynomial Euler product of (arithmetic) degree 2. Later on when referring to a normalized L-function associated with a newform we always mean $\widetilde{L}(s, f)$.

1.4.5 The Rankin-Selberg method

This is a very general and fruitful method of constructing new L-functions from given L-functions. In many situations we need sufficiently many "nice" L-functions in order to identify an automorphic object. A typical example is provided by Weil's converse theorem (Theorem 1.4.4) where many L-functions satisfying formal properties of an automorphic L-function are needed to detect a cusp form. The Rankin-Selberg method was introduced independently by

Rankin in 1939 and Selberg in 1940. The main idea is to obtain a new L-function as an integral transform of a product of given automorphic forms with "Eisenstein series" as a kernel. There are many ways to do it. Also, there are many different Eisenstein series.

Let us illustrate this idea in the case of the full modular group $\Gamma_0(1)$. An *Eisenstein series* is defined as a function of two complex variables $z \in H$ and s, $\Re s > 1$, as follows:

$$E(z, s) = \pi^{-s} \Gamma(s) \sum_{(n,m) \neq (0,0)} \frac{y^s}{|mz + n|^{2s}}.$$

It converges absolutely for $\sigma > 1$ and one easily shows that

$$E(z, s) = \pi^{-s} \Gamma(s) \zeta(s) \sum_{\gamma \in \Gamma_\infty \backslash \Gamma_0(1)} (\Im(\gamma z))^s,$$

where $\Gamma_\infty = \left\{ \begin{pmatrix} 1 & n \\ 0 & 1 \end{pmatrix} : n \in \mathbb{Z} \right\}$ is the stabilizer of the cusp at infinity. An Eisenstein series is $\Gamma_0(1)$-automorphic of weight 0 in z:

$$E(\gamma z, s) = E(z, s) \qquad (\gamma \in \Gamma_0(1)).$$

Moreover, it has meromorphic continuation to the whole s-plane with two simple poles at $s = 0$ and $s = 1$, and satisfies the following functional equation:

$$E(z, s) = E(z, 1 - s).$$

Let $f(z) = \sum_{n=1}^{\infty} a(n)e(nz)$ and $g(z) = \sum_{n=1}^{\infty} b(n)e(nz)$ be two cusp forms of weight k (in fact it suffices that only one of them is cuspidal) which are Hecke eigenfunctions, and consider

$$\Lambda(s) = \frac{1}{2} \pi^{1-k} \iint_{\Gamma_0(1)\backslash H} f(z) \overline{g(z)} E(z, s) \frac{dx\,dy}{y^2}.$$

In the case of the full modular group, the Hecke operators are self-adjoint and hence their eigenvalues are real. In particular, $b(n) \in \mathbb{R}$. Inverting the order of summation and integration and using the automorphy property we obtain

$$\Lambda(s) = 4^{1-s-k} \pi^{1-k-2s} \Gamma(s) \Gamma(s - k + 1) \zeta(2s) \sum_{n=1}^{\infty} \frac{a(n)b(n)}{n^{s+k-1}}.$$

Knowing the analytic properties of Eisenstein series, we can derive from this the corresponding properties of the *Rankin-Selberg convolution*

$$L(s, f \times g) := \zeta(2s - 2k + 2) \sum_{n=1}^{\infty} \frac{a(n)b(n)}{n^s}.$$

Theorem 1.4.5 (Rankin-Selberg) *Let f and g be Hecke eigenfunctions of weight k for the full modular group. The complete Rankin-Selberg convolution*

$$\Lambda(s, f \times g) := (2\pi)^{-2s} \, \Gamma(s) \, \Gamma(s - k + 1) \, L(s, f \times g)$$

has meromorphic continuation to the whole complex plane. It is holomorphic except for at most simple poles at $s = k$ and $s = k - 1$, and satisfies the functional equation

$$\Lambda(s, f \times g) = \Lambda(2k - 1 - s, f \times g).$$

The above theorem is just an illustration of the method which works in far more general situations. For simplicity let us state a similar result for newforms of the same weight and level, $f \in \mathcal{S}(N, k, \chi)$, $g \in \mathcal{S}(N, k, \psi)$. The corresponding normalized L-functions have Euler product expansions:

$$L(s, f) = \sum_{n=1}^{\infty} \frac{a(n)}{n^s} = \prod_p \left(1 - \frac{\alpha(p)}{p^s}\right)^{-1} \left(1 - \frac{\beta(p)}{p^s}\right)^{-1},$$

$$L(s, g) = \sum_{n=1}^{\infty} \frac{b(n)}{n^s} = \prod_p \left(1 - \frac{\gamma(p)}{p^s}\right)^{-1} \left(1 - \frac{\delta(p)}{p^s}\right)^{-1}$$

with $|\alpha(p)|$, $|\beta(p)|$, $|\gamma(p)|$, $|\delta(p)| \leqslant 1$. We define a product of these L-functions by the formula

$$L(s, f \otimes g) = \sum_{n=1}^{\infty} \frac{a(n)b(n)}{n^s}$$

and the Rankin-Selberg convolution by

$$L(s, f \times g) = L(2s, \overline{\chi\psi}) \, L(s, f \otimes g).$$

The series converges for $\sigma > 1$ and we have

$$L(s, f \otimes g) = \prod_p \left(1 - \frac{\alpha(p)\beta(p)\gamma(p)\delta(p)}{p^s}\right) \times$$

$$\prod_p \left(1 - \frac{\alpha(p)\gamma(p)}{p^s}\right)^{-1} \left(1 - \frac{\alpha(p)\delta(p)}{p^s}\right)^{-1} \left(1 - \frac{\beta(p)\gamma(p)}{p^s}\right)^{-1} \left(1 - \frac{\beta(p)\delta(p)}{p^s}\right)^{-1}.$$

Theorem 1.4.6 (Rankin-Selberg) *Let f and g be as above. Suppose that χ, ψ and $\chi\overline{\psi}$ are primitive characters $(\mathrm{mod}\, N)$. Then $L(s, f \times \overline{g})$ has meromorphic continuation to the whole complex plane and satisfies the following functional equation:*

$$\Lambda(s, f \times \overline{g}) = \epsilon \, \Lambda(1 - s, \overline{f} \times g),$$

where

$$\Lambda(s, f \times \overline{g}) = \left(\frac{(2\pi)^2}{N^{3/2}} \right)^{-s} \Gamma(s)\, \Gamma(s + k - 1)\, L(s, f \times \overline{g})$$

and $|\epsilon| = 1$.

We also have a similar result when $f = g$. In general the situation is as follows. If f and g are two newforms, not necessarily of the same type and level, then $\Lambda(s, f \times \overline{g})$ admits meromorphic continuation to $\mathbb{C}$, it is holomorphic at $s = 1$ if $f \neq g$ and has a simple pole at that point otherwise. Let $a(n, f)$ and $a(n, g)$ denote the coefficients of the normalized L-functions attached to f and g respectively. Then applying standard analytic techniques to the appropriate Rankin-Selberg convolutions, one proves the following *orthonormality property*

$$\sum_{p \leqslant x} \frac{a(p, f)\, \overline{a(p, g)}}{p} = \begin{cases} \log\log x + O(1) & \text{if } f \neq g \\ O(1) & \text{otherwise,} \end{cases} \tag{1.22}$$

the summation being over primes up to x, where $x \to \infty$.

1.4.6 Maass waves

The theory of holomorphic cusp forms can be completed by introducing Maass waves. For simplicity we focus on Maass waves of type $(0, \chi_0)$, i.e. of weight zero and principal character.

Let

$$\Delta = -y^2 \left(\frac{\partial^2}{\partial x^2} + \frac{\partial^2}{\partial y^2} \right)$$

denote the hyperbolic Laplacian on the upper half-plane. Let $L^2(\Gamma_0(N) \backslash H)$ denote the space of $\Gamma_0(N)$-invariant functions on H satisfying

$$\iint_D |f(x)|^2 \, \frac{\mathrm{d}x \, \mathrm{d}y}{y^2} < \infty$$

(D is a fundamental domain of $\Gamma_0(N)$). It is known that the Laplacian Δ acting on $L^2(\Gamma_0(N) \backslash H)$ has a continuous spectrum $[1/4, \infty)$ and a discrete spectrum

$$0 = \lambda_0 < \lambda_1 \leqslant \lambda_2 \leqslant \dots$$

Each eigenvalue has a finite multiplicity and $\lambda_j \to \infty$ as $j \to \infty$. Every $0 < \lambda_j < 1/4$ is called *exceptional*.

Selberg's Eigenvalue Conjecture *There are no exceptional eigenvalues for congruence subgroups.*

This is true for the full modular group $\Gamma_0(1)$ and for some other groups of small levels. In general the conjecture is open, but we have always $\lambda_1 \geqslant 3/16$, as was proved by A. Selberg himself.

By a *non-holomorphic cusp form (Maass wave, Maassform) of weight* 0 with respect to $\Gamma_0(N)$ we understand a smooth function f on H which

(i)   is an eigenfunction of Δ,

(ii) belongs to $L^2(\Gamma_0(N)\backslash H)$, and

(iii) vanishes at all cusps of $\Gamma_0(N)$.

Let f be a cusp form associated with an eigenvalue λ ($\Delta f = \lambda f$), and let γ_A be a scaling matrix of the cusp A. The Fourier expansion of f at the cusp A has the form

$$f(\gamma_A z) = \sqrt{y} \sum_{n \neq 0} \rho_A(n) K_{i\kappa}(2\pi |n| y) e(nx)$$

where

$$\kappa^2 = \lambda - \frac{1}{4},$$

so that κ is real or pure imaginary (if the Selberg Eigenvalue Conjecture fails). Here $K_{i\kappa}$ denotes the familiar Bessel K-function. The coefficients $\rho_A(n)$ are called the *Fourier coefficients* of f at A. When $A = i\infty$, we simply denote them by $\rho(n)$. If $\rho(-n) = \rho(n)$, we call f *even*. If $\rho(-n) = -\rho(n)$, we call f *odd*. Every Maass form is a sum of an odd and an even Maass forms, hence it is enough to study forms of a given parity.

The associated L-function is defined as the Dirichlet series

$$L(s, f) = \sum_{n=1}^{\infty} \frac{\rho(n)}{n^s}.$$

The series converges for $\sigma > 3/2$ since we have $\rho(n) \ll \sqrt{n}$.

Theorem 1.4.7 (Maass) *Let f be a Maass cusp form with eigenvalue $\lambda = 1/4 + \kappa^2$ and level N. Let*

$$\epsilon = \begin{cases} 0 & \text{if } f \text{ is even} \\ 1 & \text{if } f \text{ is odd} \end{cases}$$

and

$$\Lambda(s, f) = \left(\frac{\pi}{\sqrt{N}}\right)^{-s} \Gamma\left(\frac{s + \epsilon + i\kappa}{2}\right) \Gamma\left(\frac{s + \epsilon - i\kappa}{2}\right) L(s, f).$$

Then $\Lambda(s, f)$ extends to an entire function which is bounded on the vertical strips and satisfies the following functional equation:

$$\Lambda(s, f) = \overline{\Lambda(1 - \overline{s}, f)}.$$

The Hecke operators act on Maass waves and one has a result analogous to Theorem 1.4.1. Properties of L-functions in holomorphic and non-holomorphic case are in many aspects similar. As we shall see in the next section, both theories are special cases of a more general one. There are also differences. For instance, functional equations look differently. More importantly, the analogue of Deligne's theorem is not proved in the non-holomorphic case, and hence the Ramanujan conjecture for Maass waves is still open.

Further reading: [6], [16], [27], [29].

1.5 Representation theory and general automorphic L-functions

We start by explaining how holomorphic cusp forms can be described in terms of group representations.

The group $G := \mathrm{SL}_2(\mathbb{R})$ acts transitively on H. Hence $G \ni g \mapsto gi \in H$ is surjective and we can identify the upper half-plane with the quotient G/K, where

$$K = \left\{ R_\theta = \begin{pmatrix} \cos\theta & -\sin\theta \\ \sin\theta & \cos\theta \end{pmatrix} : 0 \leqslant \theta < 2\pi \right\}$$

denotes the stabilizer of i. Hence every $g \in G$ can be identified with (z, θ), where $z \in H$ and $\theta \in [0, 2\pi)$. On G we have the Laplacian defined as follows:

$$\Delta = -y^2 \left(\frac{\partial^2}{\partial x^2} + \frac{\partial^2}{\partial y^2} \right) - y \frac{\partial^2}{\partial x\, \partial\theta} \qquad (z = x + iy).$$

Let $\Gamma = \Gamma_0(1)$ be the full modular group and let f be a cusp form of weight k. With these data we define $\phi_f : G \to \mathbb{C}$ by the formula:

$$\phi_f(g) = (ci + d)^{-k} f(gi) \qquad \left(g = \begin{pmatrix} a & b \\ c & d \end{pmatrix} \in G \right).$$

ϕ_f has the following properties:

(1) $\phi_f(\gamma g) = \phi_f(g)$ for all $\gamma \in \Gamma$.

(2) $\phi_f(g R_\theta) = e^{-ik\theta} \phi(g)$.

(3) $\phi_f \in L^2(\Gamma \backslash G)$.

(4) For every $g \in G$ we have

$$\int_0^1 \phi_f\left(\begin{pmatrix} 1 & x \\ 0 & 1 \end{pmatrix} g \right) \mathrm{d}x = 0.$$

(5) $\Delta\phi_f = -k(k-2)\phi_f/4$.

One can prove that the converse statement is true: if a function $\phi : G \to \mathbb{C}$ satisfies (1)–(5), then there exists an $f \in \mathcal{S}(1, k, 1)$ with $\phi_f = \phi$. Hence the functions satisfying (1)–(5) and the cusp forms are essentially the same objects. Similarly we can identify the Maass waves with the functions on G satisfying (1)–(4) and such that

(5*) ϕ is an eigenfunction of Δ.

Let us observe that here we consider Maass waves of arbitrary weights, in contrast to Section 1.4.6 where the case $k = 0$ was treated. Observe also that ϕ_f can be defined for arbitrary $f \in \mathcal{S}(k, N, \chi)$. In all these cases ϕ_f has properties similar to (1)–(5). One has to introduce a factor $\chi(\gamma)$ on the right-hand side of (1) and modify suitably (4) by taking into account all the cusps of $\Gamma_0(N)$ and their "widths".

For brevity, let L_0^2 denote the subspace of $\phi \in L^2(\Gamma \backslash G)$ satisfying (4). All what we have done is just a reformulation of our previous discussion and the main issue here is that we have switched from the upper half-plane to functions defined on $\mathrm{SL}_2(\mathbb{R})$. This is important because we can now see connections with representation theory. There is one especially important representation of G, namely the *right regular representation R* in the Hilbert space of unitary operators on L_0^2 defined by the formula

$$R(g)\phi(h) = \phi(hg) \qquad (g, h \in G, \ \phi \in L_0^2).$$

The most important property of R is that it commutes with the Laplacian, i.e.,

$$\Delta R \phi = R \Delta \phi$$

for every smooth $\phi \in L_0^2$. Hence from the general theory of group representations we know that R decomposes into a direct sum of Δ-invariant subspaces, which are induced by cusp forms as we have just seen. Hence cusp forms can in fact be identified with irreducible subrepresentations of R in L_0^2. This was first observed by Gelfand and Fomin in the early 50's.

From the arithmetic viewpoint $\mathrm{SL}_2(\mathbb{R})$ is not the best group to work with. It is a completion of $\mathbb{Q}$ at the infinite prime, but there are infinitely many finite primes and they all have to be considered at the same time, if we wish to discuss Euler products for example. Therefore we change $\mathbb{R}$ to the adele ring $A_{\mathbb{Q}}$ and $\mathrm{SL}_2(\mathbb{R})$ to $G_{\mathbf{A}} := \mathrm{GL}_2(A_{\mathbb{Q}})$, which is the restricted direct product of $\mathrm{GL}_2(\mathbb{R})$ and all $\mathrm{GL}_2(\mathbb{Q}_p)$ corresponding to p-adic fields $\mathbb{Q}_p$. (The change from SL_2 to GL_2 is made for technical reasons.) Observe that $G_{\mathbf{A}}$ contains $G_{\mathbb{Q}} := \mathrm{GL}_2(\mathbb{Q})$ as a discrete subgroup.

Let $Z(G_{\mathbf{A}}) = \left\{ \begin{pmatrix} a & 0 \\ 0 & a \end{pmatrix} : a \in A_{\mathbb{Q}}^{\times} \right\}$ denote the center of $G_{\mathbf{A}}$. Now, as before, every cusp form f defines a function ϕ_f on $\mathrm{GL}_2(A_{\mathbb{Q}})$ which is $Z(G_{\mathbf{A}})G_{\mathbb{Q}}$-automorphic:

$$\phi_f(z \gamma g) = \phi_f(g) \qquad (z \in Z(G_{\mathbf{A}}), \ \gamma \in G_{\mathbb{Q}}, \ g \in G_{\mathbf{A}}).$$

Moreover, ϕ_f satisfies a cuspidal condition like (4), so that it lies in $L_0^2(Z_{\mathbf{A}} G_{\mathbb{Q}} \backslash G_{\mathbf{A}})$. As before we consider the right regular representation R of $G_{\mathbf{A}}$ on $L_0^2(Z_{\mathbf{A}} G_{\mathbb{Q}} \backslash G_{\mathbf{A}})$. Every irreducible subrepresentation π of R is called *automorphic*. Every such π has the form $\pi = \bigotimes_p \pi_p$, where for each p (including the infinite prime), π_p is an irreducible unitary representation of $\mathrm{GL}_2(\mathbb{Q}_p)$. If f is an eigenform for all the Hecke operators, $T_p(f) = a(p)f$, then

ϕ_f generates an irreducible subspace of $L_0^2(Z_{\mathbf{A}} G_{\mathbb{Q}} \backslash G_{\mathbf{A}})$. The corresponding representation $\pi = \bigotimes_p \pi_p$ is described locally in terms of $a(p)$. In particular, if f is a newform of level N and

$$\left(1 - \frac{\alpha(p)}{p^s}\right)^{-1} \left(1 - \frac{\beta(p)}{p^s}\right)^{-1}$$

is the local factor of $L(s, f)$ at p not dividing N, then π_p is represented by the matrix

$$\begin{pmatrix} \alpha(p) & 0 \\ 0 & \beta(p) \end{pmatrix}.$$

All the above theory can be generalized to the case of arbitrary reductive algebraic groups, for instance to $\mathrm{GL}_n(A_K)$, where n denotes a positive integer and K is an arbitrary number field. We shall be very sketchy and remark only that in this general context one defines cuspidal automorphic representations as certain subrepresentations of the right regular representation on a suitably defined space L_0^2. Again, every such representation is a product $\pi = \bigotimes_v \pi_v$ of local representations π_v. For almost all finite primes v, π_v is represented by a matrix of the form

$$A_v = \begin{pmatrix} \alpha_1(v) & & & \\ & \alpha_2(v) & & \\ & & \ddots & \\ & & & \alpha_n(v) \end{pmatrix}$$

which is used to define the local L-function as

$$L(s, \pi_v) = \left(\det\left(I - A_v N(v)^{-s}\right)\right)^{-1}$$

($N(v)$ denotes the norm of v). One also defines local L-functions for infinite primes and for the finite number of ramifying finite primes. Then the global *Langlands L-function* of π is defined as the product of all local L-functions:

$$L(s, \pi) = \prod_v L(s, \pi_v). \tag{1.23}$$

Theorem 1.5.1 (Godement, Jacquet, Langlands) *If π is an irreducible cuspidal automorphic representation of $\mathrm{GL}_n(A_K)$, then the product in (1.23) converges in some right half-plane. It extends to a meromorphic function on $\mathbb{C}$. When $n = 1$ and π is trivial, its only singularity is a simple pole at $s = 1$. Otherwise, $L(s, \pi)$ is entire. In either case $L(s, \pi)$ satisfies the following functional equation:*

$$L(s, \pi) = \omega_\pi L(1 - s, \widetilde{\pi}),$$

where $\widetilde{\pi}$ denotes the representation "contragradient" to π, and ω_π has absolute value 1.

This theory subsumes the theory of Hecke L-functions and the theory of L-functions attached to modular forms. The latter case was discussed at the beginning of this section. Hecke L-series with arbitrary Grössencharacters are exactly $GL_1(A_K)$ L-functions for an algebraic number field K.

Langlands' Reciprocity Conjecture *For every irreducible representation* ρ *of the Galois group* $\mathrm{Gal}(K/k)$ *there exists an automorphic representation* π *of* $GL_n(A_k)$ *such that* $L(s, K/k, \rho) = L(s, \pi)$.

Hence conjecturally automorphic L-functions contain all Artin L-functions. One can go a step further and predict that every "arithmetic" L-function is one of $L(s, \pi)$ or can be obtained from them by taking an appropriate product.

We observe that in the abelian case the Langlands reciprocity conjecture reduces to the Artin reciprocity law and therefore is known to be correct. Moreover, from Theorem 1.5.1 we see that it implies Artin's conjecture (see Section 1.3).

We close this section with some remarks about the Langlands functoriality conjecture, which goes further than his reciprocity conjecture and asserts that operations on Artin L-functions correspond to operations on automorphic representations. We cannot give details here, and we confine ourselves with vague heuristics. Suppose that ρ is a representation of a Galois group $\mathrm{Gal}(K/k)$ and let $L(s, K/k, \rho)$ be the corresponding Artin L-function. If T is a functor from the category of vector spaces to itself, then every group representation ρ induces in a natural way a representation ρ^T. The reciprocity conjecture asserts that $L(s, K/k, \rho) = L(s, \pi)$ and $L(s, K/k, \rho^T) = L(s, T(\pi))$ for certain automorphic representations π and $T(\pi)$. The functoriality conjecture predicts that the correspondence $\pi \mapsto T(\pi)$ can be canonically continued to the space of all automorphic representations.

For example take the functor of the r-th symmetric power $V \mapsto \bigvee^r V$. Suppose $\rho : \mathrm{Gal}(K/k) \to GL_2(\mathbb{C})$ is a Galois representation, and α and β are the eigenvalues of $\rho(g)$ for a certain $g \in \mathrm{Gal}(K/k)$. Then the composition of ρ and $\bigvee^r$ defines another Galois representation $\bigvee^r\rho : \mathrm{Gal}(K/k) \to GL_{r+1}(\mathbb{C})$ with eigenvalues

$$\alpha^r, \ \alpha^{r-1}\beta, \ \alpha^{r-2}\beta^2, \ \ldots, \ \beta^r.$$

Hence according to the Langlands functoriality conjecture, if π is an automorphic representation of GL_2 such that

$$L(s, \pi) = \prod_v \left(1 - \frac{\alpha(v)}{N(v)^s}\right)^{-1} \left(1 - \frac{\beta(v)}{N(v)^s}\right)^{-1}$$

then there exists an automorphic representation $\bigvee^r\pi$ such that

$$L\left(s, \bigvee^r\pi\right) = \prod_v \prod_{j=0}^{r} \left(1 - \frac{\alpha(v)^j \beta(v)^{r-j}}{N(v)^s}\right)^{-1}.$$

Similarly, considering the tensor product of two representations, the functoriality conjecture predicts that if

$$L(s, \pi_k) = \prod_v \prod_{j-1}^{n_k} \left(1 - \frac{\alpha_{k,j}(v)}{N(v)^s}\right)^{-1} \qquad (k = 1, 2),$$

where π_1 and π_2 are two automorphic representations, then there exists an automorphic representation $\pi = \pi_1 \times \pi_2$ such that

$$L(s, \pi) = \prod_v \prod_{j=1}^{n_1} \prod_{l=1}^{n_2} \left(1 - \frac{\alpha_{1,j}(v)\,\alpha_{2,l}(v)}{N(v)^s}\right)^{-1}.$$

Another example concerns the so-called *base change*. Let k, E and K be algebraic number fields, $k \subset E \subset K$, and suppose that K/k is normal, $G = \mathrm{Gal}(K/k)$, $H = \mathrm{Gal}(K/E)$. Every representation ρ of G defines by restriction a representation ρ' of H, and we have $L(s, K/E, \rho') = L(s, K/k, \rho)$, cf. Theorem 1.3.1, *3*. According to the Langlands functoriality conjecture, there should exist the corresponding operation sending automorphic representations of GL_n over k to automorphic representations over E, $\pi \mapsto B(\pi)$ (the *base change lift*), such that

$$L(s, \pi) = L(s, B(\pi)).$$

Theorem 1.5.2 (Arthur, Clozel) *Base change lift exists if E/k is cyclic.*

Further reading: [1], [4], [6], [12], [13], [17], [27].

2 The Selberg class: basic facts

2.1 Definitions and initial remarks

There is a natural desire for an axiomatic definition of the class of L-functions of number-theoretical interest. There is a certain freedom in choosing axioms, and many approaches are possible (cf. for example [7], [8], [34], [37]). One was proposed by A. Selberg [38] in 1989.

For a complex function f, let $\overline{f}$ be defined by $\overline{f}(s) = \overline{f(\overline{s})}$.

The *Selberg class* of L-functions, denoted by $\mathcal{S}$, consists of the Dirichlet series

$$F(s) = \sum_{n=1}^{\infty} \frac{a(n)}{n^s}$$

which satisfy the following axioms.

(1) The Dirichlet series converges absolutely for $\sigma > 1$.

(2) (*Analytic continuation*) There exists an integer $m \geqslant 0$ such that the function $(s-1)^m F(s)$ is entire of finite order.

(3) (*Functional equation*) F satisfies the following functional equation

$$\Phi(s) = \omega \overline{\Phi}(1-s),$$

where

$$\Phi(s) = Q^s \prod_{j=1}^{r} \Gamma(\lambda_j s + \mu_j) F(s) = \gamma(s) F(s),$$

and $r \geqslant 0$, $Q > 0$, $\lambda_j > 0$, $\Re \mu_j \geqslant 0$, $|\omega| = 1$ are parameters depending on F.

(4) (*Ramanujan hypothesis*) For every $\varepsilon > 0$ we have $a(n) \ll n^\varepsilon$.

(5) (*Euler product*) For $\sigma > 1$ we have

$$\log F(s) = \sum_{n=1}^{\infty} \frac{b(n)}{n^s},$$

where $b(n) = 0$ unless $n = p^m$ with $m \geqslant 1$, and $b(n) \ll n^\theta$ for some $\theta < 1/2$.

We stress that in the functional equation r can vanish. In this case we adopt the usual convention that an empty product equals 1, hence $\gamma(s) = Q^s$, and the functional equation takes the form

$$Q^s F(s) = \omega Q^{1-s} \overline{F}(1-s). \tag{2.1}$$

In the general case, $\gamma(s) = Q^s \prod_{j=1}^{r} \Gamma(\lambda_j s + \mu_j)$ is called the *gamma factor*. As we shall see, it is uniquely determined by $F(s)$ up to a multiplicative constant, cf. Theorem 3.1.1.

It is also convenient to introduce the *extended Selberg class* $S^\#$. It consists of $F(s)$, not identically zero, satisfying axioms (1), (2) and (3).

First of all we observe that the Selberg class contains most L-functions appearing in number theory. Here are some examples.

1. The Riemann zeta function $\zeta(s)$, see Section 1.1.

2. The shifted Dirichlet L-functions $L(s + i\theta, \chi)$, where χ is a primitive Dirichlet character $(\mathrm{mod}\, q)$, $q > 1$, and θ is a real number, see Section 1.1.

3. $\zeta_K(s)$, the Dedekind zeta function of an algebraic number field K, see Section 1.2.

4. $L_K(s, \chi)$, the Hecke L-function to a primitive Hecke character $\chi \,(\mathrm{mod}\, \mathfrak{f})$, where $\mathfrak{f}$ is an ideal of the ring of integers of K, see Section 1.2.

5. The L-function associated with a holomorphic newform of a congruence subgroup of $\mathrm{SL}_2(\mathbb{Z})$ (after suitable normalization), see Section 1.4.

6. The Rankin-Selberg convolution of any two normalized holomorphic new-forms, see Section 1.4.5.

We remark that

$F, G \in S$ implies $FG \in S$ (the same for $S^{\#}$);

If $F \in S$ is entire, then the *shift* $F_\theta(s) = F(s + i\theta)$ is in S for every real θ.

Some further functions belong to S modulo known conjectures. So we have the following list of conditional examples.

1. The Artin L-functions for irreducible representations of Galois groups (modulo Artin's conjecture: holomorphy is missing), see Section 1.3.

2. The L-functions associated with a nonholomorphic newform (the Ramanujan hypothesis is missing, exceptional eigenvalue problem), see Section 1.4.6.

3. The symmetric powers (for normalized holomorphic newforms, say): for

$$L(s) = \prod_p \left(1 - \frac{a_p}{p^s}\right)^{-1} \left(1 - \frac{b_p}{p^s}\right)^{-1},$$

the r-th symmetric power:

$$L_r(s) = \prod_p \prod_{j=0}^{r} \left(1 - a_p^j\, b_p^{r-j} p^{-s}\right)^{-1}$$

belongs to S (modulo the Langlands functoriality conjecture), see Section 1.5.

4. More generally, the $\mathrm{GL}_n(K)$ automorphic L-functions (the Ramanujan hypothesis is missing), see Section 1.5.

General examples of L-functions from the extended Selberg class are the linear combinations of solutions of the same functional equation, as for instance the Davenport-Heilbronn L-function, see Section 1.1.

We observe that owing to the Euler product condition, an L-function from the Selberg class cannot vanish in the half plane $\sigma > 1$. Moreover, using the functional equation it is easy to detect zeros on the half-plane $\sigma < 0$. They are situated at poles of the involved gamma factors and are called *trivial zeros*. The other zeros are the *non-trivial zeros*. They lie in the *critical strip*

$$0 \leqslant \sigma \leqslant 1.$$

Except for the trivial case $F \equiv 1$, there are infinitely many such zeros (this follows from (2.2) and Theorem 2.2.1 below). Trivial and non-trivial zeros can be defined in an obvious way for L-functions in the extended Selberg class. In this case, however, the critical strip is

$$1 - \sigma_F \leqslant \sigma \leqslant \sigma_F,$$

where $\sigma_F \geqslant 1$ is the least real number such that $F(s)$ does not vanish for $\sigma > \sigma_F$.

For a positive T, let $N_F(T)$ denote the number of non-trivial zeros $\rho = \beta + i\gamma$ with $|\gamma| \leqslant T$. We have the following:

$$N_F(T) = \frac{d_F}{\pi} T \log T + c_F T + O(\log T), \tag{2.2}$$

where c_F is a constant depending on F and

$$d_F = 2 \sum_{j=1}^{r} \lambda_j \tag{2.3}$$

is the *degree* of F (the empty sum equals 0). From (2.2) we see that d_F depends only on F. It is an important observation since the data λ_j, ω, Q and μ_j are not uniquely determined by F. Indeed, for instance the functional equation of the Riemann zeta function can be written as

$$\pi^{-\frac{s}{2}} \Gamma\left(\frac{s}{2}\right) \zeta(s) = \pi^{-\frac{1-s}{2}} \Gamma\left(\frac{1-s}{2}\right) \zeta(1-s)$$

and also

$$\left(\frac{\pi}{2}\right)^{-\frac{s}{2}} \Gamma\left(\frac{s}{4}\right) \Gamma\left(\frac{s}{4} + \frac{1}{2}\right) \zeta(s) = \left(\frac{\pi}{2}\right)^{-\frac{1-s}{2}} \Gamma\left(\frac{1-s}{4}\right) \Gamma\left(\frac{1-s}{4} + \frac{1}{2}\right) \zeta(1-s),$$

by the duplication formula for the Euler gamma function. Of course this is the same functional equation, but the λ_j's are different. In contrast, the right-hand side of (2.3) remains the same independently of which form of the functional equation is actually used. We shall discuss this problem in detail in Section 3.

We write

$$S_d = \{F \in S : d_F = d\} \quad \text{and} \quad S_d^{\#} = \{F \in S^{\#} : d_F = d\}.$$

Because of the importance of L-functions in number theory, we are interested in the structure of S and $S^{\#}$. It happens quite often that a property of a single L-function can be proved by considering it as a member of a family of L-functions, and hence the study of such families is important. Our main goal in these lecture notes is to discuss known results on the structure of the Selberg class. Our knowledge in this respect is very limited. We are able to find explicit description up to the degree 5/3. For larger degrees the general structural theory practically does not exist at present, except for isolated examples of L-functions. There are however quite clear expectations on what is an L-function. The most general is that they arise from automorphic representations. This is probably very hard to prove, but there are a lot of other very interesting problems to work with. Let us give some examples.

One can ask for instance about the shape of admissible functional equations. The one given in the definition of S is very general, but most of such functional equations have no solutions even in the larger class $S^{\#}$. For instance, it follows from Theorem 2.2.1 below that the functional equation

$$5^s \, \Gamma\!\left(\frac{s}{4}\right) F(s) = 5^{1-s} \, \Gamma\!\left(\frac{1-s}{4}\right) \overline{F}(1-s)$$

has no solutions in S. Much harder is to prove for instance that

$$5^s \, \Gamma\!\left(\frac{s}{2} + \sqrt{2}\,\right) F(s) = 5^{1-s} \, \Gamma\!\left(\frac{1-s}{2} + \sqrt{2}\,\right) \overline{F}(1-s)$$

has no solutions in S as well. This follows from Theorem 6.1.1 (see Section 6 below). Other forbidden classes of functional equations can be constructed using results from Section 7.

In the definition of S there are no restrictions concerning the λ_j's except that they should be real and positive. In all known instances they are rational. Even more, every known L-function satisfies a suitable functional equation with all $\lambda_j = 1/2$. Hence a good structural problem is to prove that this is always the case.

Another problem concerns Euler products. In all known cases the local factor of an L-function corresponding to a finite prime p has the form

$$\prod_{j=1}^{\partial} \left(1 - \frac{\alpha_{j,p}}{p^s}\right)^{-1}, \tag{2.4}$$

where

$$|\alpha_{j,p}| \leqslant 1 \tag{2.5}$$

and ∂ is independent of p. Moreover, for all but a finite number of primes ("ramified primes") we have

$$|\alpha_{j,p}| = 1$$

for all $j = 1, \ldots, \partial$. It is expected that every $F \in S$ has an Euler product of this type. A closely related problem is to show that ∂ (the "arithmetic degree") equals the usual ("analytic") degree d_F defined above. Another good question is to show that for every $F \in S$ there exists an integer q_F, a "conductor", such that the ramified primes are exactly the prime divisors of q_F. A good candidate for q_F will be proposed in Section 3.3.

There are hundreds of such questions, and equally many possibilities for speculations.

2.2 The simplest converse theorems

In order to illustrate what we mean by the description of the structure of S and $S^{\#}$, let us consider the following simple case of degrees less than 1.

Theorem 2.2.1 ([9]) $S_0 = \{1\}$ and $S_d = \varnothing$ if $0 < d < 1$.

Similarly one can deal with the extended Selberg class. The functional equation of $F \in S_0^\#$ has no Γ-factors (cf. (2.1)):

$$Q^s F(s) = \omega\, Q^{1-s}\, \overline{F}(1-s).$$

We call $q = q_F = Q^2$ the *conductor* of $F \in S_0^\#$. Let $S_0^\#(q, \omega)$ be the set of $F \in S_0^\#$ with given q and ω. Moreover, let

$$V_0^\#(q, \omega) = S_0^\#(q, \omega) \cup \{0\}.$$

Theorem 2.2.2 ([20]) *If $F \in S_0^\#$ then*

1. *q is a positive integer.*
2. *q and ω are uniquely determined by F, and hence $S_0^\#$ is the disjoint union of the subclasses $S_0^\#(q, \omega)$ with $q \in N$ and $|\omega| = 1$.*
3. *For such q and ω, every $F \in S_0^\#(q, \omega)$ has the form*

    $$F(s) = \sum_{n|q} a(n) n^{-s}.$$

4. *$V_0^\#(q, \omega)$ is a real vector space of dimension $d(q)$ (the divisor function).*

Theorem 2.2.3 ([20]) *We have $S_d^\# = \varnothing$ for $0 < d < 1$.*

Proofs are not complicated. From (2.1) we see that $F \in S_0^\#$ is entire since otherwise it would have a pole at $s = 0$. For real x let us consider

$$
\begin{aligned}
h(x) &= \sum_{n=1}^{\infty} a(n) e^{-2\pi n x} \\
&= \frac{1}{2\pi i} \int_{2-i\infty}^{2+i\infty} F(s) \Gamma(s) (2\pi x)^{-s} \, \mathrm{d}s \\
&= \sum_{k=0}^{\infty} \frac{(-1)^k F(-k)}{k!} (2\pi x)^k \\
&= \omega Q \sum_{k=0}^{\infty} \frac{(-1)^k Q^{2k} \overline{F}(1+k)}{k!} (2\pi x)^k.
\end{aligned}
$$

The last series converges everywhere and hence h has analytic continuation to the whole complex plane and is entire. For complex z, let

$$H(z) = h(-iz) = \sum_{k=1}^{\infty} a(n) e(nz).$$

This is a power series in $e(z) = e^{2\pi i z}$. Hence it converges everywhere and consequently

$$a(n) \ll R^{-n}$$

for every $R > 0$. In particular the Dirichlet series of F converges (absolutely) everywhere. The functional equation (2.1) can be viewed as an identity between absolutely convergent Dirichlet series

$$\sum_{n=1}^{\infty} a(n) \left(\frac{Q^2}{n} \right)^s = \omega Q \sum_{n=1}^{\infty} \frac{\overline{a(n)}}{n} n^s.$$

We use the uniqueness theorem for generalized Dirichlet series. It follows that $q = Q^2$ is an integer and $a(n) = 0$ unless $n \mid Q^2$. In particular we have that $F(s)$ is a Dirichlet polynomial supported on the divisors of the conductor:

$$F(s) = \sum_{n \mid q} a(n) n^{-s}, \tag{2.6}$$

and the functional equation takes the form

$$a(n) = \frac{\omega n}{\sqrt{q}} \overline{a(q/n)} \qquad (n \mid q). \tag{2.7}$$

Therefore q and ω are uniquely determined by F. Using (2.6) and (2.7) it is easy to compute the dimension of $V_0^{\#}(q, \omega)$. It equals $d(q)$ and Theorem 2.2.2 follows.

Proof of Theorems 2.2.1 and 2.2.3 Let $F \in S_0$. By the Euler product axiom we have

$$\log F(s) = \sum_{p \mid q} \sum_{m=1}^{\infty} \frac{b(p^m)}{p^{ms}},$$

and the series converges absolutely for $\sigma > \theta$ with $\theta < 1/2$. By the functional equation this means that F is entire and non-vanishing. Hence using Hadamard's theory we have

$$F(s) = A \cdot B^s$$

for certain constants A and B. By the functional equation (2.7) this easily implies that $q = 1$ and therefore $F \equiv 1$. The first part of Theorem 2.2.1 follows and the case of degree zero is established.

Let now $F \in S_d^{\#}$ with $0 < d < 1$. As before, let us consider for $x > 0$

$$h(x) = \sum_{n=1}^{\infty} a(n) e^{-2\pi n x}.$$

This time we cannot a priori exclude that F has a pole at $s = 1$. Therefore using the integral representation of $h(x)$ and shifting the line of integration we have

$$h(x) = \frac{P(\log x)}{x} + \sum_{k=0}^{\infty} \frac{(-1)^k F(-k)}{k!} (2\pi x)^k$$

for a certain polynomial P. By $d < 1$ and by Stirling's formula for the gamma function, using the functional equation we obtain

$$\frac{F(-k)}{k!} \ll k^{-(1-d)k} A^k$$

for some $A > 0$. Hence the series converges everywhere. This gives the analytic continuation of h to the whole complex plane with the slit along the negative real axis. Since h is periodic with period i, we conclude that the polynomial P is identically zero and h is entire. As before, this means that the Dirichlet series for F converges absolutely everywhere. In particular

$$F(\sigma + it) \ll 1$$

for every fixed σ. This is impossible since by the functional equation we have

$$|F(-\sigma + it)| \gg |t|^{d(1/2+\sigma)}$$

for sufficiently large but fixed σ and for $|t| \to \infty$. Consequently, $S_d^{\#} = S_d = \varnothing$ for $0 < d < 1$ and both theorems are proved.

2.3 Euler product

Let us now look at the Euler product axiom in the definition of the Selberg class. We have

$$\log F(s) = \sum_p \sum_{m=1}^{\infty} b(p^m) p^{-ms}.$$

For sufficiently large σ we can perform formal manipulations on the absolutely convergent series, thus obtaining

$$F(s) = \prod_p F_p(s),$$

where

$$F_p(s) = \sum_{m=0}^{\infty} a(p^m) p^{-ms}. \tag{2.8}$$

Hence the coefficients $a(n)$ are multiplicative. The product converges for $\sigma > 1$. We call F_p the *local factor of* F. The series in (2.8) converges absolutely for $\sigma > 0$. Since $b(p^m) \ll p^{m\theta}$ with $\theta < 1/2$, the series

$$\log F_p(s) = \sum_{m=0}^{\infty} b(p^m) p^{-ms}$$

converges for $\sigma > \theta$. In particular $F_p(s) \neq 0$ for $\sigma > \theta$.

In all known examples of L-functions, $1/F_p(s)$ is a polynomial in p^{-s}, so that the local factors are of the form (2.4), (2.5). If this happens for an $F \in S$ we say that F has a *polynomial Euler product*. We denote by S^{poly} the subclass of such $F \in S$.

Suppose $F, G \in S$ are different. It may happen that local factors of F and G are the same for certain primes p:

$$F_p = G_p \quad \text{for} \quad p \in \Omega_{F,G}.$$

Problem *How large can $\Omega_{F,G}$ be?*

The following simple example shows that it can be pretty large. Let χ_1 and χ_2 be two different primitive Dirichlet characters $(\mathrm{mod}\, q)$ such that $\chi_1(a) = \chi_2(a)$ for certain a coprime to q. Then the corresponding set of primes Ω contains all $p \equiv a \,(\mathrm{mod}\, q)$. In particular it can have positive density.

On the other hand there is a well-known result in representation theory called the *Strong Multiplicity One Theorem* (Piatetski-Shapiro [35]).

Theorem 2.3.1 ([35]) *Let*

$$\pi = \bigotimes_p \pi_p, \qquad \pi' = \bigotimes_p \pi'_p$$

be two automorphic representations. If $\pi_p = \pi'_p$ for almost all primes p, then $\pi = \pi'$.

Here and in the sequel "almost all" means "all but finitely many".

In the language of L-functions, Theorem 2.3.1 says that if the local factors $L_p(s, \pi)$ and $L_p(s, \pi')$ are the same for all but finitely many primes, then the global L-functions are the same. In this formulation the multiplicity one theorem holds in the Selberg class, as was proved by M. R. Murty and V. K. Murty.

Theorem 2.3.2 ([31]) *Let $F, G \in S$. If $F_p(s) = G_p(s)$ for almost all primes p, then $F = G$.*

Proof Let $F_p = G_p$ for $p \notin T$, for a certain finite set T of primes. Let us consider the quotient

$$\frac{\gamma_F(s)F(s)}{\gamma_G(s)G(s)} = \frac{\gamma_F(s)}{\gamma_G(s)} \prod_{p \in T} \frac{F_p(s)}{G_p(s)}.$$

The right-hand side is holomorphic and non-vanishing for $\sigma \geqslant 1/2$ and hence by the functional equation we have that $F(s)/G(s)$ is entire, non-vanishing and of finite order. Therefore by Hadamard's theory

$$\frac{F(s)}{G(s)} = e^{as+b} \frac{\gamma_G(s)}{\gamma_F(s)}.$$

for certain complex numbers a and b. By Stirling's formula we obtain

$$\frac{F(2+it)}{G(2+it)} = c\,e^{\alpha t}t^{\beta}e^{i\gamma t\log t}e^{i\delta t}\bigl(1+O(1/t)\bigr) \qquad (t\to\infty)$$

for certain real α, β, γ, δ and complex c. The left-hand side is almost periodic in t and hence $\alpha = \beta = 0$. Now we use the following well-known theorem of Bohr ([3]).

Theorem 2.3.3 ([3]) *If $f(t)$ is almost periodic and satisfies $|f(t)| \geqslant k > 0$, then $\arg f(t) = \lambda t + \phi(t)$ with λ real and $\phi(t)$ almost periodic.*

This implies that $\gamma = 0$ as well. Hence

$$e^{-i\delta t}\frac{F(2+it)}{G(2+it)} = c + o(1)$$

as $t \to \infty$. But the left-hand side is almost periodic and hence it has to be constant. By analytic continuation we obtain

$$e^{\delta(2-s)}F(s) = c\,G(s)$$

for all complex s. By the uniqueness theorem for generalized Dirichlet series we see that $\delta = 0$. Moreover, since $a_F(1) = a_G(1) = 1$, we have $c = 1$ and the result follows.

We observe that the above arguments show that the assumption $F_p(s) = G_p(s)$ can be replaced by the weaker requirement that

$$a_F(p) = a_G(p) \quad \text{and} \quad a_F(p^2) = a_G(p^2)$$

for almost all primes. It would be desirable to remove the condition involving squares. This is not known in general, but can be proved for certain important subclasses of S (cf. [25], [32], [41]).

Theorem 2.3.4 ([25]) *Let $F, G \in S^{\mathrm{poly}}$. If $a_F(p) = a_G(p)$ for almost all p, then $F = G$.*

The proof is based on a lemma concerning general Euler products. Let

$$f(s) = \sum_n a(n)n^{-s}, \qquad \log f(s) = \sum_n b(n)n^{-s},$$

where

$$b(n) = 0 \text{ unless } n = p^m \text{ with } m \geqslant 1, \quad b(n) \ll n^{\theta} \ (\theta < 1/2),$$

and let

$$\mu_f = \liminf_{\sigma\to 1^+}(\sigma-1)\sum_p \frac{|a(p)|^2\log p}{p^{\sigma}}.$$

Lemma 2.3.1 *Assume that $f(s)$ has meromorphic continuation to $\sigma \geqslant 1$ and that $\mu_f < \infty$. Then $f(s)$ has finitely many zeros and poles on the line $\sigma = 1$.*

For the proof see [25].

Proof of Theorem 2.3.4 Let $F, G \in S^{\mathrm{poly}}$ and $a_F(p) = a_G(p)$ for $p \notin T$, for a certain finite set T. Let $H(s) = F(s)/G(s)$. We have

$$H(s) = \prod_{p \notin T} \left(1 - \frac{c(p^2)}{p^{2s}}\right)^{-1} P(s),$$

where

$$c(p^2) = a_F(p^2) - a_G(p^2) \ll 1 \tag{2.9}$$

and $P(s)$ is holomorphic and non-vanishing for $\sigma \geqslant 1/2$. The function

$$f(s) = \prod_{p \notin T} \left(1 - \frac{c(p^2)}{p^{s}}\right)^{-1}$$

satisfies the assumptions of Lemma 2.3.1. In particular $\mu_f < \infty$ because of (2.9). Hence H has at most finitely many zeros and poles on the line $\sigma = 1/2$. We denote them (counting multiplicities) by

$$\frac{1}{2} + \imath t_j, \qquad j = 1, \ldots, K.$$

We write

$$h_j(s) = s - \frac{1}{2} - i t_j, \qquad j = 1, \ldots, K.$$

Then the h_j's satisfy functional equations of the form

$$h_j(s) = -\overline{h_j}(1 - s).$$

Hence there exists a rational function $R(s)$, satisfying

$$R(s) = \eta \overline{R}(1 - s)$$

with $\eta = \pm 1$, such that $R(s)H(s)$ is holomorphic and non-vanishing for $\sigma \geqslant 1/2$. Now, by the functional equations of F and G, we have that

$$K(s) = R(s)H(s)\frac{\gamma_F(s)}{\gamma_G(s)}$$

is holomorphic, non-vanishing for $\sigma \geqslant 1/2$ and satisfies

$$K(s) = \omega \overline{K}(1 - s)$$

for a certain ω with $|\omega| = 1$. Hence by Hadamard's theory we have $K(s) = e^{as+b}$ for certain complex constants a and b. Consequently

$$\frac{F(s)}{G(s)} = \frac{e^{as+b}}{R(s)} \frac{\gamma_G(s)}{\gamma_F(s)}.$$

By Stirling's formula we obtain

$$\frac{F(2+it)}{G(2+it)} = c\, e^{\alpha t} t^{\beta} e^{i\gamma t \log t} e^{i\delta t} \big(1 + O(1/t)\big) \qquad (t \to \infty)$$

for certain real α, β, γ, δ and complex c. Hence the result follows arguing as in the proof of Theorem 2.3.2.

2.4 Factorization

We say that $F \in S$ is *primitive* if $F = F_1 F_2$, $F_1, F_2 \in S$, implies $F_1 = 1$ or $F_2 = 1$. Similarly, $F \in S^{\#}$ is *primitive* if $F = F_1 F_2$, $F_1, F_2 \in S^{\#}$ implies $F_1 = \text{const.}$ or $F_2 = \text{const.}$ Note that the constants are the only invertible elements in $S^{\#}$.

Formally, both definitions are very similar. However, connections between primitivity in S and $S^{\#}$ are not clear. In particular it is an open problem whether a primitive function in S is primitive in $S^{\#}$ as well.

Theorem 2.4.1 ([9]) *Every $F \in S$ can be factored into primitive factors.*

Proof This follows by induction on the degree using the following facts:

1. The degree is additive: $d_{FG} = d_F + d_G$.
2. $S_0 = \{1\}$.
3. $S_d = \emptyset$ for $0 < d < 1$.

Theorem 2.4.2 ([26]) *Every $F \in S^{\#}$ can be factored into primitive factors.*

The proof is more involved, and induction on the degree does not suffice. In fact such an induction implies that every $F \in S^{\#}$ can be written as a product of almost primitive factors:

$$F(s) = F_1(s) \dots F_k(s). \tag{2.10}$$

Here and in the sequel we say that $G \in S^{\#}$ is *almost primitive* if

$$G(s) = G_1(s) G_2(s), \qquad G_1, G_2 \in S^{\#}$$

implies $d_{G_1} = 0$ or $d_{G_2} = 0$.

We need the following

Theorem 2.4.3 ([26]) *If $F \in S^{\#}$ is almost primitive then $F(s) = G(s)P(s)$ with $G, P \in S^{\#}$, $d_G = 0$, and $P(s)$ primitive.*

We shall for a moment postpone the proof of this theorem. Let us see how it implies Theorem 2.4.2. Using (2.10) we write

$$F(s) = G(s)P_1(s)\ldots P_k(s),$$

where $d_G = 0$ and P_j, $j = 1,\ldots,k$, are primitive. Since the functions in $S_0^{\#}$ have integer conductors (cf. Theorem 2.2.2) and those with conductor 1 are constants, an induction on the conductor shows that G is a product of primitive functions. Therefore Theorem 2.4.2 follows.

For the proof of Theorem 2.4.3 we need some lemmas.

Let $\delta_0 \geqslant 1$. We define $a(n,\delta_0)$, $n \geqslant 1$, by induction:

$$a(1,\delta_0) = \delta_0,$$

$$a(n,\delta_0) = \delta_0 + \sum_{l=2}^{\Omega(n)} \frac{1}{l} \sum_{\substack{n_1\ldots n_l=n \\ n_j \geqslant 2}} a(n_1,\delta_0)\ldots a(n_l,\delta_0),$$

where, as usual, $\Omega(n)$ denotes the number of prime factors of n counting multiplicities.

Lemma 2.4.1 ([26]) *For $n \geqslant 1$*

$$\delta_0 \leqslant a(n,\delta_0) \leqslant \delta_0^{\Omega(n)} 2^{\Omega(n)^3}.$$

Proof By induction (see [26]).

For a given $F \in S^{\#}$ let σ_F denote the abscissa of non-vanishing of F, and recall that q_F denotes the conductor of $F \in S^{\#}$.

Lemma 2.4.2 ([26]) *Let $F \in S_0^{\#}$ have $a(1) = 1$ and $q_F \geqslant 2$. Then*

$$M := \max_{n|q_F} |a(n)| = O_{\sigma_F}\left(e^{\log^3 q_F}\right).$$

Proof For sufficiently large σ we have

$$\log F(s) = \sum_{n=1}^{\infty} b(n)n^{-s}. \tag{2.11}$$

We may assume that $\sigma_F > 1$. We estimate $\log F(s)$ for $\sigma > \sigma_F + \delta$, for (small) positive δ. For $\sigma > \sigma_F + \delta/2$ we have $F(s) \ll_\delta M$, whence

$$\Re \log F(s) = \log|F(s)| \leqslant c_1(\delta) \log M.$$

For $\sigma > c_2(\varepsilon)\log M$ we have $F(s) = 1 + O(\varepsilon)$. Hence by the Borel-Carathéodory theorem $\log F(s) = O_\delta(\log^2 M)$ for $\sigma > \sigma_F + \delta$, and the Lindelöf μ-function of $\log F(s)$ satisfies $\mu(\sigma) = 0$ for $\sigma > \sigma_F$. Consequently the series in (2.11) converges for $\sigma > \sigma_F$ and therefore it converges absolutely for $\sigma > \sigma_F + 1$. For such σ we have

$$b(n)n^{-\sigma} = \lim_{T\to\infty} \frac{1}{2T} \int_{-T}^{T} \log F(\sigma + it)n^{it}dt \ll \log^2 M.$$

Therefore

$$|b(n)| \leqslant \delta_0\, n^\sigma \log^2 M$$

for some $\delta_0 > 0$ and every $\sigma > \sigma_F + 1$.

We now express the coefficients $b(n)$ in terms of the coefficients $a(n)$. For sufficiently large σ we have $F(s) = 1 + G(s)$, say, where $|G(s)| < 1/2$. Hence

$$\log F(s) = \log(1 + G(s)) = \sum_{l=1}^{\infty} \frac{(-1)^{l+1}}{l} G(s)^l$$

$$= \sum_{l=1}^{\infty} \frac{(-1)^{l+1}}{l} \sum_{n=1}^{\infty} n^{-s} \sum_{\substack{n_1,\dots,n_l \geqslant 2 \\ n_1 \dots n_l = n}} a(n_1)\dots a(n_l).$$

Comparing the Dirichlet coefficients we obtain

$$b(n) = a(n) + \sum_{l=2}^{\Omega(n)} \frac{(-1)^{l+1}}{l} \sum_{\substack{n_1,\dots,n_l \geqslant 2 \\ n_1 \dots n_l = n}} a(n_1)\dots a(n_l).$$

By induction we get from this

$$|a(n)| \leqslant n^\sigma a(n, \delta_0) \log^{2\Omega(n)} M, \qquad (\sigma > \sigma_F + 1),$$

where $a(n, \delta_0)$ is the sequence defined before Lemma 2.4.1. Consequently

$$M \leqslant q_F^\sigma \max_{n|q_F} \left(a(n, \delta_0) \log^{2\Omega(n)} M \right). \tag{2.12}$$

We consider two cases.

Case I: $M \leqslant \exp(\log^3 q_F)$. In this case our lemma is obvious.

Case II: $M > \exp(\log^3 q_F)$. Since $\Omega(n) \leqslant \dfrac{\log x}{\log 2}$, using (2.12) and Lemma 2.4.1 we have

$$M \ll q_F^\sigma M^{1/2} \delta_0^{2(\log q_F)/\log 2} e^{4\log^3 q_F}$$

whence putting $\sigma = \sigma_F + 2$ we obtain

$$M \ll q_F^{2\sigma} \delta_0^{\log q_F} e^{8\log^3 q_F} \ll_{\sigma_F} e^{10\log^3 q_F}$$

as required.

For $F \in S^\#$ let

$$N_F(T) = \#\{\rho : F(\rho) = 0,\ |\beta| \leqslant \sigma_F,\ |\gamma| < T\}$$

be the usual zero-counting function.

Lemma 2.4.3 ([26]) *We have*

$$N_F(T) = \frac{T}{\pi} \log q_F + O_{\sigma_F}(\log^6 q_F)$$

uniformly for $T \geqslant 2$ and $F \in S_0^{\#}$ with $a(1) = 1$ and $q_F \geqslant 2$.

Proof Let $\sigma_0 > \sigma_F$ be such that

$$|F(s) - 1| \leqslant \frac{1}{4} \quad \text{for} \quad \sigma \geqslant \sigma_0.$$

σ_0 exists since $a(1) = 1$. Let R be the rectangle with vertices $\sigma_0 \pm iT$ and $1 - \sigma_0 \pm iT$, where T is not the ordinate of a zero of F. Then by standard arguments

$$N_F(T) = \frac{1}{2\pi} \Delta_{\partial R} \arg(Q^s F(s)) = \frac{T}{\pi} \log q_F + O(\sigma_0 \log(q_F^\varepsilon M)).$$

Now choosing $\sigma_0 = c \log^3 q_F$ and using Lemma 2.4.2, we conclude the proof.

Proof of Theorem 2.4.3 Let $F \in S^{\#}$ be almost primitive. If F is not primitive we can write

$$F(s) = L_1(s) F_1(s)$$

for certain almost primitive F_1 and $L_1 \in S_0^{\#}$ with $q_{L_1} \geqslant 2$. If F_1 is not primitive we apply inductively the same reasoning, and hence arguing by contradiction we may assume that for every positive integer n

$$F(s) = L_1(s) \dots L_n(s) F_n(s)$$

with F_n almost primitive and $L_j \in S_0^{\#}$, $q_{L_j} \geqslant 2$ for $j = 1, \dots, n$. Looking at the Dirichlet series on both sides we have that only a finite number of L_j's can have the first coefficient $a_{L_j}(1) = 0$. Therefore, by a normalization, for sufficiently large n we can write

$$F(s) = H(s) H_1(s) \dots H_n(s) F_n(s)$$

with $d_H = 0$, $d_{H_j} = 0$, $q_{H_j} \geqslant 2$, $a_{H_j}(1) = 1$ $(j = 1, \dots, n)$, and F_n almost primitive. Writing $G_n = H_1 \dots H_n$ we finally obtain

$$F(s) = H(s) G_n(s) F_n(s)$$

with $d_H = 0$, $d_{G_n} = 0$, $q_{G_n} \to \infty$ as $n \to \infty$, $a_{G_n}(1) = 1$ and F_n almost primitive.

Note that

$$N_F(T) \geqslant N_{G_n}(T) \quad \text{and} \quad \sigma_{G_n} \leqslant \sigma_F$$

for all n. Hence by Lemma 2.4.3 we have

$$\frac{d_F}{\pi} T \log T \geqslant \frac{T}{2\pi} \log q_{G_n} + O\left(\log^6 q_{G_n}\right).$$

Choosing $T = q_{G_n}^{\delta}$ with a sufficiently small positive δ we get a contradiction and Theorem 2.4.3 follows.

2.5 Selberg conjectures

In these lecture notes we are mainly interested in unconditional results. Nevertheless it is interesting to discuss briefly some spectacular consequences of the following Selberg Orthonormality Conjecture (SOC for short), motivated by the properties of the Rankin-Selberg convolution (cf. (1.22)).

Conjecture (SOC) *For every primitive $F, G \in S$ we have*

$$\sum_{p \leqslant x} \frac{a_F(p)\, \overline{a_G(p)}}{p} = \delta_{F,G} \log \log x + O(1)$$

as $x \to \infty$, where

$$\delta_{F,G} = \begin{cases} 1 & \text{if } F = G, \\ 0 & \text{if } F \neq G. \end{cases}$$

Theorem 2.5.1 ([9], [30]) *Assume SOC. Then the factorization into primitive functions in S is unique up to the order of factors.*

Proof Let $P_1, \ldots, P_m, Q_1, \ldots, Q_n$ be distinct primitive elements of the Selberg class such that

$$P_1^{e_1}(s) \ldots P_m^{e_m}(s) = Q_1^{f_1}(s) \ldots Q_n^{f_n}(s)$$

for certain positive integers $e_1, \ldots, e_m$ and $f_1, \ldots, f_n$. Comparing p-th coefficients on both sides we have

$$e_1 a_{P_1}(p) + \ldots + e_m a_{P_m}(p) = f_1 a_{Q_1}(p) + \ldots + f_n a_{Q_n}(p).$$

Multiplying by $\overline{a_{P_1(p)}}/p$ and summing over primes $p \leqslant x$ we obtain

$$e_1 \sum_{p \leqslant x} \frac{|a_{P_1}(p)|^2}{p} + \sum_{j=2}^{m} e_j \sum_{p \leqslant x} \frac{a_{P_j}(p)\overline{a_{P_1}(p)}}{p} = \sum_{j=1}^{n} f_j \sum_{p \leqslant x} \frac{a_{Q_j}(p)\overline{a_{P_1}(p)}}{p}.$$

By SOC, the left hand side is $e_1 \log \log x + O(1)$ whereas the right hand side is $O(1)$, a contradiction.

If $F = P_1^{e_1} \ldots P_m^{e_m}$ is a factorization into powers of distinct primitive functions, then SOC implies that

$$\sum_{p\leqslant x} \frac{|a_F(p)|^2}{p} = n_F \log\log x + O(1),$$

where n_F is an integer given by the formula

$$n_F = \sum_{i=1}^{m} e_i^2.$$

Hence, under SOC, F is primitive if and only if $n_F = 1$.

Theorem 2.5.2 ([9]) *Assume SOC. Then the Riemann zeta function is the only polar and primitive element of S.*

Proof If $F \in S$ has a pole of order $m > 0$ at $s = 1$ then

$$\sum_p \frac{a(p)}{p^\sigma} = \log F(\sigma) + O(1) = m \log(\sigma - 1) + O(1)$$

as $\sigma \to 1^+$. Suppose that F is primitive and different from $\zeta(s)$. Then by SOC the sum

$$S_F(x) = \sum_{p\leqslant x} \frac{a(p)}{p}$$

is bounded as $x \to \infty$. Therefore

$$\sum_p \frac{a(p)}{p^\sigma} = (\sigma - 1) \int_1^\infty S_F(x) x^{-\sigma} \mathrm{d}x$$

is bounded as $\sigma \to 1^+$, a contradiction. Hence the result follows.

As an immediate consequence we obtain that SOC implies the *Dedekind conjecture*: $\zeta_{\mathbb{Q}} \mid \zeta_K$ for any algebraic number field K. Indeed, $\zeta_K(s)$ decomposes uniquely into primitive factors and at least one of these factors has a pole at $s = 1$. Hence it equals the Riemann zeta function. The same argument, by Landau's theorem, shows the following more general result.

Theorem 2.5.3 *Assume SOC. Then every $F \in S$ with non-negative Dirichlet coefficients is divisible by the Riemann zeta function.*

Theorem 2.5.4 ([30]) *SOC implies Artin's conjecture: every Artin L-function $L(s, \rho, K/k)$ associated with a non-trivial irreducible representation ρ of the Galois group $\mathrm{Gal}(K/k)$ has an analytic continuation to the whole complex plane. Moreover, $L(s, \rho, K/k)$ is primitive in S.*

Proof Using Theorem 1.3.1 it suffices to consider the case of absolute Galois extensions. Hence without loss of generality we assume that $k = \mathbb{Q}$. As explained in Section 1.3, $L(s, \rho, K/\mathbb{Q})$ can be written as a quotient of products of Hecke L-functions of some intermediate fields $\mathbb{Q} \subset E_j \subset K$:

$$L(s, \rho, K/k) = \frac{\displaystyle\prod_{j=1}^{m} L_{E_j}(s, \chi_j)}{\displaystyle\prod_{j=m+1}^{n} L_{E_j}(s, \chi_j)}.$$

After factoring each Hecke L-function into primitive factors in S we obtain

$$L(s, \rho, K/\mathbb{Q}) = \prod_{j=1}^{N} F_j^{e_j}(s)$$

where e_j, $j = 1, \ldots, N$, are integers and F_j are primitive and distinct.

Let χ denote the character of ρ and σ_p the Frobenius automorphism associated with the prime p. By SOC we can write

$$\sum_{p \leqslant x} \frac{|\chi(\sigma_p)|^2}{p} = \left(\sum_{j=1}^{N} e_j^2\right) \log \log x + O(1).$$

On the other hand, using Chebotarev's density theorem one can check that

$$\sum_{p \leqslant x} \frac{|\chi(\sigma_p)|^2}{p} = \log \log x + O(1).$$

Therefore

$$\sum_{j=1}^{N} e_j^2 = 1$$

so that $N = 1$ and $e_1 = \pm 1$. Since $L(s, \rho, K/k)$ has trivial zeros, the case $e_1 = -1$ is excluded and the result follows.

Theorem 2.5.5 ([30]) *SOC implies the Langlands reciprocity conjecture for solvable extensions $K/\mathbb{Q}$.*

Proof First we observe that Jacquet and Shalika [18], [19] and Shahidi [40] established analytic properties of the Rankin-Selberg convolution of two GL_n automorphic L-functions, which suffice to verify that for every irreducible automorphic representation π the Dirichlet coefficients $a(n, \pi)$ of the associated L-function $L(s, \pi)$ satisfy

$$\sum_{p \leqslant x} \frac{|a(p, \pi)|^2}{p} = \log \log x + O(1).$$

Hence, assuming SOC, $L(s, \pi)$ is primitive in S provided it is a member of the Selberg class, i.e. if the Ramanujan conjecture holds for this function.

Let K be a Galois extension of $\mathbb{Q}$ of degree n with solvable Galois group G. Hence there exists a chain of consecutive Galois extensions of prime degrees

$$\mathbb{Q} \subset K_1 \subset \ldots \subset K_n = K.$$

Using repeatedly Theorem 1.5.2, we see that the Dedekind zeta function is an automorphic L-function, and therefore

$$\zeta_K(s) = \prod_{i=1}^{r} L(s, \pi_i)^{e_i}, \tag{2.13}$$

where π_i are irreducible cuspidal automorphic representations over $\mathbb{Q}$, and e_i are positive integers. Comparing local Euler factors on both sides, we see that the Ramanujan conjecture holds for all $L(s, \pi_i)$. Hence, according to our initial remark, (2.13) is a factorization into primitive elements in S. On the other hand by (1.12) we have

$$\zeta_K(s) = \prod_{\chi} L(s, \chi, K/\mathbb{Q})^{\dim \chi}, \tag{2.14}$$

where the product is over the irreducible characters of G. According to Theorem 2.5.4, this is a factorization into primitive elements in S. By unique factorization (Theorem 2.5.1), (2.13) and (2.14) coincide up to the order of factors, and hence each $L(s, K/\mathbb{Q}, \chi)$ is one of $L(s, \pi_i)$. The reciprocity conjecture therefore follows.

3 Functional equation and invariants

3.1 Uniqueness of the functional equation

Theorem 3.1.1 ([9]) *If γ_1 and γ_2 are gamma factors for $F \in S^{\#}$, then $\gamma_1(s) = C\gamma_2(s)$ for some constant $C \neq 0$.*

Proof Let us consider the quotient of two functional equations of F with gamma factors $\gamma_1(s)$ and $\gamma_2(s)$ respectively. We have

$$h(s) = \omega \overline{h}(1 - s),$$

where

$$h(s) = \gamma_1(s)/\gamma_2(s)$$

and ω is a complex number with $|\omega| = 1$. We know that h is holomorphic and non-vanishing for $\sigma > 0$. Hence h is entire and non-vanishing. Moreover, by Stirling's formula it is of order 1 or less. Hence

$$h(s) = e^{as+b}$$

for certain complex a and b. By the functional equation of h we have

$$e^{as+b} = \omega e^{\overline{a}(1-s)+\overline{b}}.$$

Hence $a = i\alpha$ for a certain real α. Putting $s = it$, $t \to \infty$, we have by Stirling's formula

$$\left| \frac{h(it)}{h(-it)} \right| \to 1.$$

Therefore $\alpha = 0$ and the result follows.

3.2 Transformation formulae

The following problem arises. We know that the gamma factors of $F \in S^{\#}$ are uniquely determined up to a multiplicative constant. Nevertheless they can have different shapes due to identities involving the Euler gamma function, such as the *Legendre-Gauss multiplication formula*

$$\Gamma(s) = m^{s-1/2}(2\pi)^{(1-m)/2} \prod_{k=0}^{m-1} \Gamma\left(\frac{s+k}{m}\right)$$

or the *factorial formula*

$$\Gamma(s+1) = s\Gamma(s).$$

What are the admissible forms of the gamma factors of a given $F \in S^{\#}$?

Theorem 3.2.1 ([21]) *Let $\gamma_1(s)$ and $\gamma_2(s)$ be two gamma factors of $F \in S^{\#}$, $\gamma_1 = C\gamma_2$. Then γ_1 can be transformed into $C\gamma_2$ by repeated applications of the Legendre-Gauss multiplication formula and the factorial formula.*

Instead of giving a formal proof of this theorem we confine ourselves to the enlightening special case of

$$\gamma(s) = Q^s \Gamma(\lambda s + \mu), \qquad Q > 0, \quad \lambda > 0 \text{ and } \Re\mu \geqslant 0.$$

Let l_j and m_j be integers with $0 \leqslant l_j < m_j$. A family (l_j, m_j), $j = 1, \ldots, M$, is called an *exact covering system* if for every integer k there exists a unique $1 \leqslant j \leqslant M$ such that $k \equiv l_j \pmod{m_j}$. It is clear that every exact covering system satisfies

$$\sum_{j=1}^{M} \frac{1}{m_j} = 1.$$

Moreover, writing $L = \text{l.c.m.}(m_1, \ldots, m_M)$ we have

$$\left\{ k_j m_j + l_j : j = 1, \ldots, M, \ k_j = 0, \ldots, \frac{L}{m_j} - 1 \right\} = \{0, \ldots, L-1\}.$$

Proposition 3.2.1 ([21])

(1) *If the identity*

$$\Gamma(s) = e^{as+b} \prod_{j=1}^{M} \Gamma(\alpha_j s + \beta_j) \tag{3.1}$$

holds, then there exists an exact covering system (l_j, m_j), $j = 1, \ldots, M$, *such that*

$$\alpha_j = \frac{1}{m_j}, \quad \beta_j = \frac{l_j}{m_j}. \tag{3.2}$$

(2) *For every exact covering system* (l_j, m_j), $j = 1, \ldots, M$, *there exist complex numbers* a, b *such that* (3.1) *and* (3.2) *hold.*

(3) *Every identity of type* (3.1) *can be obtained starting with* $\Gamma(s)$ *and applying the Legendre-Gauss multiplication formula at most* $(M+1)$ *times.*

Proof Let us compare poles on both sides of (3.1). The poles on the left hand side are simple and located at non-positive integers, whereas those on the right hand side are at

$$s = -\frac{\beta_j + l}{\alpha_j}, \quad l \geqslant 0, \quad 1 \leqslant j \leqslant M.$$

Therefore

$$\frac{\beta_j}{\alpha_j}, \quad \frac{\beta_j + 1}{\alpha_j}$$

are non-negative integers. Consequently

$$m_j := \frac{1}{\alpha_j} \quad \text{and} \quad l_j := m_j \beta_j = \frac{\beta_j}{\alpha_j}$$

are integers, and (3.2) holds. To show that (l_j, m_j), $j = 1, \ldots, M$, is an exact covering system we write (3.1) in the form

$$\Gamma(s) = e^{as+b} \prod_{j=1}^{M} \Gamma\left(\frac{s + l_j}{m_j}\right).$$

Note that for every non-negative integers k and l there exists a unique $1 \leqslant j \leqslant M$ such that $\dfrac{-k + l_j}{m_j} = -l$. Hence $k \equiv l_j \pmod{m_j}$. Now we show that $l_j \leqslant m_j - 1$ for $j = 1, \ldots, M$ by contradiction. Suppose that $l_{j_0} \geqslant m_{j_0}$ for some j_0 and consider the residue $l_{j_0}^* \equiv l_{j_0} \pmod{m_{j_0}}$, $0 \leqslant l_{j_0}^* < m_{j_0}$. It is clear that $s = -l_{j_0}^*$ is not a pole of $\Gamma\left(\dfrac{s + l_{j_0}}{m_{j_0}}\right)$ and hence there exists $(l_{j_1}, m_{j_1}) \neq (l_{j_0}, m_{j_0})$ such that $l_{j_0}^* \equiv l_{j_1} \pmod{m_{j_1}}$. But then the integer $k = l_{j_0}^* + m_{j_0} m_{j_1}$ satisfies

$$k \equiv l_{j_0} \pmod{m_{j_0}} \quad \text{and} \quad k \equiv l_{j_1} \pmod{m_{j_1}},$$

a contradiction. Hence (l_j, m_j), $j = 1, \ldots, M$, is an exact covering system and the first part of the theorem follows.

To show (2) let us consider

$$f(s) = \frac{\Gamma(s)}{\displaystyle\prod_{j=1}^{M} \Gamma\left(\frac{s + l_j}{m_j}\right)}.$$

This function is entire and non-vanishing. Moreover it is of order 1 or less. Hence $f(s) = e^{as+b}$ as required.

We show (3). According to (1), (3.1) has the form

$$\Gamma(s) = e^{as+b} \prod_{j=1}^{M} \Gamma\left(\frac{1}{m_j}s + \frac{l_j}{m_j}\right) \tag{3.3}$$

where (l_j, m_j), $j = 1, \ldots, M$, is an exact covering system. Let $L :=$ l.c.m.$(m_1, \ldots, m_M)$. We apply the Legendre-Gauss multiplication formula:

$$\Gamma(s) = e^{a_0 s + b_0} \prod_{k=0}^{L} \Gamma\left(\frac{s + k}{L}\right).$$

Then we split the factors according to the residue classes. The product equals

$$e^{a_0 s + b_0} \prod_{j=1}^{M} \prod_{k_j=0}^{\frac{L}{m_j}-1} \Gamma\left(\frac{s + k_j m_j + l_j}{L}\right) =$$

$$e^{a_0 s + b_0} \prod_{j=1}^{M} \prod_{k_j=0}^{\frac{L}{m_j}-1} \Gamma\left(\frac{(s + l_j)/m_j + k_j}{L/m_j}\right).$$

We apply the multiplication formula with $m = L/m_j$. The inner product equals

$$e^{a_j s + b_j} \Gamma\left(\frac{s + l_j}{m_j}\right).$$

Hence we obtain (3.3) with

$$e^{as+b} = \prod_{j=1}^{M} e^{a_j s + b_j},$$

and the result follows.

The general case is technically more involved but in principle very similar. First we split the product

$$Q^s \prod_{j=1}^{r} \Gamma(\lambda_j s + \mu_j)$$

into $\mathbb{Q}$-equivalence classes, λ_{j_1} and λ_{j_2} being $\mathbb{Q}$-equivalent when their quotient is rational. Then we apply the above procedure (suitably generalized) to each equivalence class. Since Γ-functions belonging to different equivalence classes can have common poles, the use of the factorial formula is necessary. Here is an example illustrating this situation:

$$\Gamma(s)\,\Gamma(\sqrt{2}s + 1) = \sqrt{\frac{2}{\pi}}\, 2^s \, \Gamma\left(\frac{s}{2} + 1\right) \Gamma\left(\frac{s+1}{2}\right) \Gamma(\sqrt{2}s).$$

As an application of Proposition 3.2.1 we obtain the following complete description of admissible forms of gamma factors in case of Dirichlet L-functions. Of course other known L-functions can be treated similarly.

Theorem 3.2.2 ([21]) *Let* $\chi\,(mod\,q)$ *be a primitive Dirichlet character. Then all the γ-factors of $L(s,\chi)$ are of the form*

$$Q^s \prod_{j=1}^{M} \Gamma\left(\frac{s}{2m_j} + \frac{2l_j + a(\chi)}{2m_j}\right),$$

where (l_j, m_j), $j = 1, \ldots, M$, is any exact covering system,

$$Q = \left(\frac{q}{\pi} \prod_{j=1}^{M} m_j^{1/m_j}\right)^{1/2}$$

and $a(\chi) = (1 + \chi(-1))/2$.

3.3 Invariants

Let $F \in S^{\#}$ have the gamma factor of the usual form

$$\gamma(s) = Q^s \prod_{j=1}^{r} \Gamma(\lambda_j s + \mu_j). \tag{3.4}$$

Instead of giving a formal definition of a parameter and an invariant of F, we restrict ourselves to the following definition. We call any function

$$p(Q, \lambda_1, \ldots, \lambda_r, \mu_1, \ldots, \mu_r, \omega),$$

depending on the data Q, λ_j, μ_j and ω, a *parameter* of F. If a parameter depends only on F and is independent of the particular form of the gamma factor, then it is called an *invariant* of F. As an immediate consequence of Theorem 3.2.1 we have the following result.

Theorem 3.3.1 ([21]) *A parameter is an invariant of $F \in S^{\#}$ with $d_F > 0$ if and only if it is stable by the multiplication formula and the factorial formula.*

The following examples of important invariants explain what we mean by this statement.

(I) *H-invariants*

For a non-negative integer n let

$$H_F(n) = 2 \sum_{j=1}^{r} \frac{B_n(\mu_j)}{\lambda_j^{n-1}}, \tag{3.5}$$

where $B_n(x)$ denotes the n-th Bernoulli polynomial:

$$\frac{ze^{zx}}{e^z - 1} = \sum_{n=0}^{\infty} B_n(x) \frac{z^n}{n!} \qquad (|z| < 2\pi).$$

We have

$$B_0(x) = 1, \quad B_1(x) = x - \frac{1}{2}, \quad B_2(x) = x^2 - x + \frac{1}{6}, \ \dots$$

whence

$$H_F(0) = 2 \sum_{j=1}^{r} \lambda_j = d_F \qquad \text{(the degree)},$$

$$H_F(1) = 2 \sum_{j=1}^{r} \left(\mu_j - \frac{1}{2} \right) = \xi_F = \eta_F + i\theta_F \qquad \text{(the ξ-invariant).}$$

We check the invariance of $H_F(n)$ using Theorem 3.3.1. We have the following identities for the Bernoulli polynomials:

$$B_n(x+1) = B_n(x) + nx^{n-1} \qquad (n \geqslant 0) \tag{3.6}$$

and

$$B_n(mx) = m^{n-1} \sum_{j=0}^{m-1} B_n\left(x + \frac{j}{m} \right) \qquad (n \geqslant 0, \ m \geqslant 1). \tag{3.7}$$

By applying the multiplication formula to a Γ-factor

$$\Gamma(\lambda s + \mu),$$

it becomes

$$e^{as+b} \prod_{j=0}^{m} \Gamma\left(\frac{\lambda s + \mu + j}{m}\right)$$

with suitable a and b. Thus the parameters change according to the following rules:

$$\lambda \mapsto \lambda_j := \frac{\lambda}{m} \qquad (j = 0, \ldots, m-1),$$

$$\mu \mapsto \mu_j := \frac{\mu + j}{m} \qquad (j = 0, \ldots, m-1).$$

Hence a summand

$$\frac{B_n(\mu)}{\lambda^{n-1}}$$

in the definition of $H_F(n)$ is replaced by

$$\sum_{j=0}^{m} \frac{B_n\left(\frac{\mu}{m} + \frac{j}{m}\right)}{(\lambda/m)^{n-1}},$$

whence the sum on the right hand side of (3.5) remains unchanged by (3.7). This means that $H_F(n)$ is stable by the multiplication formula.

We pass to the factorial formula. We always apply it to a pair of Γ-factors

$$\Gamma(\lambda s + \mu)\,\Gamma(\lambda' s + \mu') = (\lambda s + \mu - 1)\,\Gamma(\lambda s + \mu - 1)\,\Gamma(\lambda' s + \mu')$$

$$= \frac{\lambda}{\lambda'}\left(\lambda' s + \frac{(\mu - 1)\lambda'}{\lambda}\right)\Gamma(\lambda s + \mu - 1)\,\Gamma(\lambda' s + \mu').$$

In order to absorb the linear factor we need the following consistency condition:

$$\frac{\mu - 1}{\lambda} = \frac{\mu'}{\lambda'}. \tag{3.8}$$

Then our expression becomes

$$\frac{\lambda}{\lambda'}\,\Gamma(\lambda s + \mu - 1)\,\Gamma(\lambda' s + \mu' + 1).$$

Thus in order to prove that $H_F(n)$ is stable by the factorial formula we have to check that

$$\frac{B_n(\mu)}{\lambda^{n-1}} + \frac{B_n(\mu')}{\lambda'^{n-1}} = \frac{B_n(\mu - 1)}{\lambda^{n-1}} + \frac{B_n(\mu' + 1)}{\lambda'^{n-1}}.$$

But this follows from (3.6) under the consistency condition (3.8).

(II) *The conductor*

We have introduced the conductor for *L*-functions of degree 0 in Section 2.2: $q_F = Q^2$. For positive degrees we put

$$q_F = (2\pi)^{d_F} Q^2 \prod_{j=1}^{r} \lambda_j^{2\lambda_j}.$$

Again it is easy to check that the conductor is stable by the multiplication and factorial formulae, and hence q_F is an invariant.

(III) *The root number*

We define it by the following formula:

$$\omega_F^* = \omega e^{-i\frac{\pi}{2}(\eta_F+1)} \left(\frac{q}{(2\pi)^{d_F}} \right)^{i\theta_F/d_F} \prod_{j=1}^{r} \lambda_j^{-2i\Im\mu_j}.$$

It is easy to check the invariance of the root number.

H-invariants, conductor and root number form a set of "basic" invariants. The exact meaning of this statement is explained by the following theorem.

Theorem 3.3.2 ([22]) $F, G \in S^{\#}$ *satisfy the same functional equation if and only if $q_F = q_G$, $\omega_F^* = \omega_G^*$ and $H_F(n) = H_G(n)$ for every $n \geqslant 0$.*

Proof For an $F \in S^{\#}$ let

$$K_F(z) := -z \sum_{\rho} e^{-\rho z} \qquad (\Re z > 0),$$

where the summation is over all poles of $\gamma_F(s)$. If $\gamma_F(s)$ is as in (3.4), the generic ρ equals

$$-\frac{\mu_j + k}{\lambda_j}, \qquad j = 1, \ldots, r, \quad k \geqslant 0.$$

Hence

$$K_F(z) = -z \sum_{j=1}^{r} \sum_{k=0}^{\infty} e^{\frac{\mu_j+k}{\lambda_j} z}$$

$$= \sum_{j=1}^{r} \frac{z e^{\mu_j z/\lambda_j}}{e^{z/\lambda_j} - 1}.$$

Thus for $|z| < 2\pi \min |\lambda_j|$ we have

$$K_F(z) = \sum_{j=1}^{r} \lambda_j \sum_{n=0}^{\infty} B_n(\mu_j) \frac{z^n}{n! \lambda_j^n}$$

$$= \frac{1}{2} \sum_{n=0}^{\infty} \frac{H_F(n)}{n!} z^n.$$

Since $K_F(z)$ is clearly an invariant of F, we obtain another proof of the invariance of the $H_F(n)$'s. Now, if $H_F(n) = H_G(n)$ for every $n \geqslant 0$ then $K_F(z) = K_G(z)$, and by the uniqueness property of the generalized Dirichlet series we have that $\gamma_F(s)$ and $\gamma_G(s)$ have the same poles. Therefore

$$\gamma_F(s) = e^{as+b} \gamma_G(s) \tag{3.9}$$

for certain complex a and b. Using Stirling's formula for $t \to \pm\infty$ and $\sigma = 0$ we see that a is real. Hence the functional equation for F can be written as follows:

$$e^{as} \gamma_G(s) F(s) = \left(\omega e^{\overline{b}-b} \right) e^{a(1-s)} \overline{\gamma_G}(1-s) \overline{F}(1-s). \tag{3.10}$$

Computing the conductor of F using (3.10) we obtain

$$q_F = e^{2a} q_G = e^{2a} q_F,$$

whence $a = 0$. Consequently (3.9) takes the form

$$\gamma_F(s) = c\, \gamma_G(s)$$

for a certain complex c. Since we have $\omega_F^* = \omega_G^*$, c has to be real and the result follows.

The general project would be to describe the structure of the Selberg class by describing admissible values of the invariants. We have therefore the following general problem.

Problem *Given an invariant $I : S \to \mathbb{C}$ or $I : S^{\#} \to \mathbb{C}$, describe $I(S)$ or $I(S^{\#})$.*

For some invariants we have good conjectures about their sets of values. The following two special cases are particularly important.

Degree Conjecture $d_F \in \mathbb{N} \cup \{0\}$ *for every $F \in S^{\#}$.*

Conductor Conjecture $q_F \in \mathbb{N}$ *for every $F \in S$.*

Here are examples of degrees and conductors of some classical L-functions.

DEGREES AND CONDUCTORS OF SOME L-FUNCTIONS

F	d_F	q_F
Riemann $\zeta(s)$	1	1
Dirichlet $L(s,\chi)$, $\chi \,(\mathrm{mod}\, q)$ primitive	1	q
Dedekind $\zeta_K(s)$	$[K:\mathbb{Q}]$	$|D_K|$
Hecke $L_K(s,\chi)$, $\chi \,(\mathrm{mod}\,\mathfrak{f})$ primitive	$[K:\mathbb{Q}]$	$N(\mathfrak{f})|D_K|$
$L(s,f)$, $f \in S(N,k,\chi)$ newform	2	N
Artin $L(s,K/k,\chi)$	$[k:\mathbb{Q}]\dim\chi$	$|D_k|N_{k/\mathbb{Q}}(\mathfrak{f}(\chi,K/k))$

4 Hypergeometric functions

4.1 Gauss hypergeometric function

For a complex number a and a non-negative integer k we write as usual

$$(a)_k = \begin{cases} 1 & \text{if } k = 0 \\ a(a+1)\ldots(a+k-1) & \text{if } k \geqslant 1. \end{cases}$$

The Gauss hypergeometric function is defined by

$$F(a,b,c,z) = \sum_{k=0}^{\infty} \frac{(a)_k\,(b)_k}{(c)_k\,k!}\, z^k.$$

The hypergeometric series converges absolutely and uniformly on compact subsets of $|z| < 1$, and hence $F(a,b,c,z)$ is holomorphic for $|z| < 1$ (cf. [11]). Moreover, $F(a,b,c,z)$ has analytic continuation to $\mathbb{C} \setminus [1,\infty)$ as a single-valued holomorphic function.

Lemma 4.1.1 ([20]) *Let $c \neq 0, -1, \ldots$, $c-a-b \neq 1, 2, \ldots$, $\Re(c-a-b) > 0$, and $\rho > 0$. Then, uniformly for $\phi \neq 0$, we have*

$$\lim_{\rho \to 0^+} F(a,\, b,\, c,\, 1 + \rho e^{i\phi}) = \frac{\Gamma(c)\,\Gamma(c-a-b)}{\Gamma(c-a)\,\Gamma(c-b)}.$$

4.2 Complete and incomplete Fox hypergeometric functions

In the general case the Fox hypergeometric function is defined by the Barnes-Mellin integral. The kernel function is defined as a quotient of products of Γ-functions. We fix parameters

$$p, q, n, m \text{ integers}, \ 0 \leqslant n \leqslant p, \ 0 \leqslant m \leqslant q,$$

$$\alpha_j \ (j = 1, \ldots, p), \ \beta_j \ (j = 1, \ldots, q) \text{ positive numbers},$$

$$a_j \ (j = 1, \ldots, p), \ b_j \ (j = 1, \ldots, q) \text{ complex numbers},$$

$$\alpha_k(b_h + \nu) \neq \beta_h(a_k - 1 - \lambda), \ \nu, \lambda \geqslant 0, \ h = 1, \ldots, m, \ k = 1, \ldots, n.$$

With these data we build a kernel function as follows:

$$h(w) = \frac{\displaystyle\prod_{j=1}^{n} \Gamma(1 - a_j + \alpha_j w) \prod_{j=1}^{m} \Gamma(b_j - \beta_j w)}{\displaystyle\prod_{j=m+1}^{q} \Gamma(1 - b_j + \beta_j w) \prod_{j=n+1}^{p} \Gamma(a_j - \alpha_j w)}.$$

The *Fox hypergeometric function* is defined by

$$H(z) = \frac{1}{2\pi i} \int_C h(w) z^w \, \mathrm{d}w$$

where C denotes a path from $-i\infty$ to $+i\infty$ such that the points

$$s = \frac{b_j + \nu}{\beta_j} \qquad (1 \leqslant j \leqslant m, \ \nu \geqslant 0)$$

lie to the right of C, and the points

$$s = \frac{a_j - 1 - \nu}{\alpha_j} \qquad (1 \leqslant j \leqslant n, \ \nu \geqslant 0)$$

to the left of C.

When C is a path running from $-i\infty$ to $+i\infty$ but in a different way, we speak of *incomplete Fox hypergeometric function*. Of course the complete Fox function and incomplete Fox function with the same kernel function differ by the sum of the corresponding residues.

The *main parameter* of the Fox function is defined as follows:

$$\mu = \sum_{j=1}^{q} \beta_j - \sum_{j=1}^{p} \alpha_j. \tag{4.1}$$

4.3 The first special case: $\mu = 0$

We need a special case of incomplete Fox function. We have data

$$\lambda_1, \ldots, \lambda_r > 0, \qquad d := 2\sum_{j=1}^{r} \lambda_j, \qquad \mu_1, \ldots, \mu_r \text{ complex.}$$

For a positive integer K let

$$H_K(z, s) = H_K(z, s; \boldsymbol{\lambda}, \boldsymbol{\mu}) = \frac{1}{2\pi i} \int_{(-K-1/2)} h(w, s) z^w \, dw$$

where

$$h(w, s) = h^*(w, s) \, \Gamma(s)$$

and

$$h^*(w, s) = \prod_{j=1}^{r} \frac{\Gamma\left(\lambda_j(1-s) + \overline{\mu_j} - \frac{\lambda_j}{d}w\right)}{\Gamma\left(\lambda_j s + \mu_j + \frac{\lambda_j}{d}w\right)}.$$

Hence in this case the parameters are:

$$n = 1, \quad p = 1, \quad m = r, \quad q = 2r,$$

$$\alpha_1 = 1, \quad a_1 = 1,$$

$$\beta_j = \frac{\lambda_j}{d}, \quad b_j = \lambda_j(1-s) + \overline{\mu_j} \quad (j = 1, \ldots, r),$$

$$\beta_j = \frac{\lambda_{j-r}}{d}, \quad b_j = 1 - \lambda_{j-r}s - \mu_{j-r} \quad (j = r+1, \ldots, 2r).$$

Here the main parameter (cf. (4.1)) is $\mu = 0$.

We observe that our parameters are not constants but functions of s. Hence in our case the Fox function depends on two complex variables. We need to know what happens as $|z| \to \infty$ or $z \to -i/\beta$, with β defined in (4.2) below, whereas s stays in a compact set.

We first observe that we can restrict ourselves to the case $d = 1$. This is due to the following transformation formula

$$H_K(z, s; \lambda_1, \ldots, \lambda_r, \mu_1, \ldots, \mu_r) = H_K(z, \tilde{s}; \tilde{\lambda}_1, \ldots, \tilde{\lambda}_r, \tilde{\mu}_1, \ldots, \tilde{\mu}_r),$$

where

$$\tilde{s} = ds - \frac{d-1}{2}, \quad \tilde{\lambda}_j = \frac{\lambda_j}{d}, \quad \tilde{\mu}_j = \mu_j + \frac{\lambda_j}{2}\left(1 - \frac{1}{d}\right)$$

for $j = 1, \ldots, r$. Indeed, we have

$$\tilde{d} = 2\sum_{j=1}^{r} \tilde{\lambda}_j = 1.$$

Write

$$\beta = \prod_{j=1}^{r} \lambda_j^{-2\lambda_j} \tag{4.2}$$

and consider the following subsets of the complex plane:

$$A = \{z \in \mathbb{C} : \Re z > 0\},$$

$$B_\beta = \left\{z \in \mathbb{C} : |z| < \frac{1}{\beta}\right\} \setminus \left(-\frac{1}{\beta}, \, 0\right],$$

$$C_\beta = \left\{z \in \mathbb{C} : |z| > \frac{1}{\beta}\right\},$$

$$D_\beta = A \cup B_\beta \cup C_\beta.$$

Let R be a positive real number. For

$$K := \left[\max_{1 \leqslant j \leqslant r} \frac{1 + 2|\mu_j|}{2\lambda_j} + 2 \sum_{j=1}^{r} \left(|\mu_j| + \frac{1}{2}\right) + R \right] + 1$$

we have the following result.

Lemma 4.3.1 ([20], [24]) *The integral*

$$\frac{1}{2\pi i} \int_{(-K-1/2)} h(w, s) z^w \mathrm{d}w$$

is absolutely and uniformly convergent on compact subsets of $A \times \{\sigma < R\}$ and $H_K(z, s)$ has holomorphic continuation to $D_\beta \times \{\sigma < R\}$ as a single-valued function. For real y, $|y| \to \infty$, and complex s satisfying $-L \leqslant \sigma \leqslant R$, $L, R > 0$, we have

$$H_K(iy, s) \ll (|t| + 2)^{cL} |y|^{-K-1/2}.$$

The proof is based on a detailed inspection of the asymptotic behaviour of the kernel function. Since it is quite technical and long, we refer the interested reader to the original papers [20] and [24].

Crucial for us is the behaviour of the Fox function when $y = -1/\beta$. We write

$$H_K(-i/\beta, s) = \lim_{z \to -i/\beta} H_K(z, s).$$

Lemma 4.3.2 ([24]) *The function $H_K(-i/\beta, s)$ is meromorphic for $\sigma < R$ with at most simple poles at the points*

$$s^{(k)} = 1 - k - i\theta, \quad k \geqslant 0,$$

and non-vanishing residue at $s^{(0)}$.

We give a brief sketch of the proof, leaving the details to the reader (see [20] and [24]). Let

$$A(s) = \sum_{j=1}^{r} (\lambda_j(1-2s) - 2i\Im\mu_j) \log 2\lambda_j, \quad a = \frac{1}{2}(1-s) + \frac{1}{2}\bar{\xi}, \quad b = \frac{1}{2}s + \frac{1}{2}\xi.$$

Using Stirling's formula we obtain

$$h^*(w,s) = e^{A(s)} \frac{\Gamma\left(a + \frac{1}{2} - \frac{1}{2}w\right)}{\Gamma\left(b + \frac{1}{2} + \frac{1}{2}w\right)} (\beta/2)^w (1 + f(w,s)),$$

where

$$f(w,s) = O\left(1/|w|\right).$$

Thus, after the change of variable $\frac{1}{2}\beta z \mapsto z$, the function $H_K(z,s)$ is approximated by

$$H_K^*(z,s) := \frac{e^{A(s)}}{2\pi i} \int_{(-K-1/2)} \frac{\Gamma\left(a + \frac{1}{2} - \frac{1}{2}w\right)}{\Gamma\left(b + \frac{1}{2} + \frac{1}{2}w\right)} \Gamma(w) z^w \, dw$$

$$= e^{A(s)} \Bigg(-\sum_{k=0}^{K} \frac{(-1)^k}{k!} \frac{\Gamma\left(a + \frac{1}{2} + \frac{1}{2}k\right)}{\Gamma\left(b + \frac{1}{2} - \frac{1}{2}k\right)} z^{-k}$$

$$+ \frac{\cos(\pi b)}{\pi} \Gamma\left(a + \frac{1}{2}\right) \Gamma\left(\frac{1}{2} - b\right) F\left(a + \frac{1}{2}, \frac{1}{2} - b, \frac{1}{2}, -\frac{1}{4z^2}\right)$$

$$- \frac{\sin(\pi b)}{\pi z} \Gamma(a+1) \Gamma(1-b) F\left(a+1, 1-b, \frac{3}{2}, -\frac{1}{4z^2}\right) \Bigg),$$

where $F(A, B, C, z)$ denotes the classical Gauss hypergeometric function. Since the β-parameter of the Fox function $e^{-A(s)} H_K^*(z,s)$ equals 2, using Lemma 4.1.1 we have

$$e^{-A(s)} H_K^*(-i/2, s) = -\sum_{k=0}^{K} \frac{1}{k!} \frac{\Gamma\left(a + \frac{1}{2} + \frac{1}{2}k\right)}{\Gamma\left(b + \frac{1}{2} - \frac{1}{2}k\right)} \left(\frac{i}{2}\right)^{-k}$$

$$+ \frac{\cos(\pi b)}{\sqrt{\pi}} \frac{\Gamma\left(a + \frac{1}{2}\right) \Gamma\left(\frac{1}{2} - b\right) \Gamma\left(b - a - \frac{1}{2}\right)}{\Gamma(-a) \Gamma(b)}$$

$$- i \frac{\sin(\pi b)}{\sqrt{\pi}} \frac{\Gamma(a+1) \Gamma(1-b) \Gamma\left(b - a - \frac{1}{2}\right)}{\Gamma\left(\frac{1}{2} - a\right) \Gamma\left(\frac{1}{2} + b\right)}.$$

This gives meromorphic continuation to $\sigma > 0$. Using more terms in Stirling's formula we obtain meromorphic continuation to the whole complex plane. The expression for $H_K^*(-i/2, s)$ is sufficiently explicit to study the polar structure. This is somewhat involved, but one finally proves that there are at most simple poles and that they are at points $s^{(k)}$, $k \geqslant 0$. The residue at $s^{(0)}$ does not vanish and hence there is really a pole at that point.

4.4 The second special case: $\mu > 0$

In Section 7 we shall need the Fox hypergeometric function built from the data of the γ-factor of $F \in S_d^\#$, with $d > 1$, in the following way:

$$n = 1, \quad p = 1, \quad m = r, \quad q = 2r,$$

$$\alpha_1 = 1, \quad a_1 = 1,$$

$$\beta_j = \lambda_j, \quad b_j = \lambda_j(1 - s) + \overline{\mu_j} \quad (j = 1, \ldots, r),$$

$$\beta_j = \lambda_{j-r}, \quad b_j = 1 - \lambda_{j-r}s - \mu_{j-r} \quad (j = r+1, \ldots, 2r).$$

In this case we have the incomplete Fox function

$$H_c(z, s) = \frac{1}{2\pi i} \int_{(-c)} h(w, s) z^w dw,$$

where

$$h(w, s) = \prod_{j=1}^r \frac{\Gamma(\lambda_j(1 - s - w) + \overline{\mu_j})}{\Gamma(\lambda_j(s + w) + \mu_j)} \Gamma(w).$$

Here s lies in a strip $1 - L < \sigma < 2$ with some large L, $c = c(s)$ is real and $\sigma < c < 2$, $c \neq 0, 1$. Moreover $z = x + 2\pi i \alpha$, $x > 0$, α real. We need information on $H_c(z, s)$ when z is purely imaginary, and hence we are going to take $x \to 0^+$.

The main parameter of $H_c(z, s)$ (cf. (4.1)) is $\mu = d - 1 > 0$, since $d > 1$. Using the identity

$$\Gamma(z)\Gamma(1 - z) = \frac{\pi}{\sin \pi z}$$

we write $h(w, s)$ as

$$h(w, s) = h_0(w, s)h_1(w, s)$$

with

$$h_0(w, s) = \frac{1}{\Gamma(1 - w)} \prod_{j=1}^r \Gamma(\lambda_j(1 - s - w) + \overline{\mu_j}) \Gamma(1 - \lambda_j(s + w) - \mu_j)$$

and

$$h_1(w, s) = \frac{\pi^{1-r}}{\sin \pi w} \prod_{j=1}^r \sin\left(\pi(\lambda_j(s + w) + \mu_j)\right).$$

Lemma 4.4.1 ([23]) *Let* $1 < d < 2$, *and let* J *be an integer. Moreover, let*

$$\theta = 2\Im \sum_{j=1}^r \left(\mu_j - \frac{1}{2}\right), \quad a = \beta(d - 1)^{d-1} \quad \text{and} \quad b = \beta(d - 1)^d,$$

where β *is given by* (4.2). *Then there exist a constant* $c_0 \neq 0$ *and polynomials* $p_j(s)$, *with* $0 \leqslant j \leqslant J$ *and* $p_0(s) = c_0$ *identically, such that*

$$h_0(w,s) = a^w b^s \sum_{j=0}^{J} p_j(s)\, \Gamma\Big((1-d)w - \frac{d}{2} - ds - i\theta - j\Big)$$

$$+ a^w b^s\, \Gamma\Big((1-d)w - \frac{d}{2} - ds - i\theta\Big) f_0(w,s),$$

where

$$f_0(w,s) \ll |\Im w|^{-J-1}, \qquad |\Im w| \to \infty,$$

uniformly for s in any compact subset of $1 - L < \sigma < 2$.

The proof is quite technical and we skip it (see [23]).

Lemma 4.4.2 ([23]) *Let d, L, σ and w be as above. Moreover, let*

$$\xi = 2 \sum_{j=1}^{r} \Big(\mu_j - \frac{1}{2}\Big).$$

Then there exists a constant $0 < \lambda < d$ such that

$$h_1(w,s) = -\frac{2\pi i}{(2\pi)^r}\, e^{-i\pi \frac{d}{2} s - i\frac{\pi}{2}\xi}\, e^{i\pi(1-d/2)w} + f_1(w,s),$$

where

$$f_1(w,s) \ll \begin{cases} e^{-\pi(1-\lambda/2)\Im w} & \text{if } \Im w \to +\infty \\ e^{\pi(1-d/2)|\Im w|} & \text{if } \Im w \to -\infty \end{cases}$$

uniformly for s in any compact subset of $1 - L < \sigma < 2$.

The proof is easy, see [23].

According to the above lemmas we split $H_c(z,s)$ as follows:

$$H_c(z,s) = \sum_{j=0}^{J} H_c^{(j)}(z,s) + \widetilde{H}_c^{(1)}(z,s) + \widetilde{H}_c^{(2)}(z,s),$$

where

$$H_c^{(j)}(z,s) = -\frac{b^s p_j(s) e^{-i\pi \frac{d}{2} s - i\frac{\pi}{2}\xi}}{(2\pi)^r} \times$$

$$\int_{(-c)} \Gamma\Big((1-d)w + \frac{d}{2} - ds - i\theta - j\Big)\big(a e^{i\pi(1-d/2)} z\big)^w dw,$$

$$\widetilde{H}_c^{(1)}(z,s) = \frac{b^s}{2\pi i} \sum_{j=0}^{J} p_j(s) \times$$

$$\int_{(-c)} \Gamma\Big((1-d)w + \frac{d}{2} - ds - i\theta - j\Big) f_1(w,s)(az)^w dw,$$

$$\widetilde{H}_c^{(2)}(z,s) = \frac{b^s}{2\pi i} \times$$

$$\int_{(-c)} \Gamma\Big((1-d)w + \frac{d}{2} - ds - i\theta\Big) f_0(w,s) h_1(w,s)(az)^w dw.$$

By Lemmas 4.4.1 and 4.4.2, the integrals in $\widetilde{H}_c^{(j)}$, $j = 1, 2$, converge absolutely and uniformly for $|\arg z| \leqslant \pi/2$ and s in a compact subset of $1 - L < \sigma < 2$. Hence we can pass to the limit for $x \to 0^+$ in these cases. The integrals in $H_c^{(j)}$, $0 \leqslant j \leqslant J$, can be explicitly computed. We formulate the result in a form suitable for future applications in Section 7.

Lemma 4.4.3 ([23]) *Let*

$$\kappa = \frac{1}{d-1}, \quad A = (d-1)q^{-\kappa}, \quad B = b\Big(\frac{a}{2\pi Q^2}\Big)^{-d\kappa},$$

and

$$L_j = -\frac{2\pi i e^{-i\frac{\pi}{2}\xi}}{(2\pi)^r (d-1)} \Big(\frac{ae^{i\pi(1-d/2)}}{2\pi i Q^2}\Big)^{\frac{d/2 - i\theta - j}{d-1}}, \quad D_k = \frac{(-1)^k}{k!} \Big(\frac{ae^{i\pi(1-d/2)}}{2\pi i Q^2}\Big)^{k\kappa}.$$

Then we have

$$H_c^{(j)}\Big(-i\frac{n}{2\pi Q^2 \alpha}, s\Big) =$$

$$L_j p_j(s) B^s \Big(\frac{\alpha}{n}\Big)^{\frac{ds - d/2 + i\theta + j}{d-1}} \Big\{ e\Big(A\Big(\frac{n}{\alpha}\Big)^\kappa\Big) - \sum_{k=0}^{j} D_k \psi_{j-k}(s)\Big(\frac{\alpha}{n}\Big)^{-k\kappa} \Big\}$$

where

$$\psi_l(s) = \begin{cases} 1 & \text{if } d/2 + (c-\sigma)(d-1) < \sigma < d/2 + (c-\sigma)(d-1) + 1 \\ 0 & \text{if } (d-1)\sigma_a(F) - J + 1 < \sigma < d/2 + (c-\sigma)(d-1) - l \end{cases}$$

uniformly for s in a compact subset of the strip

$$d/2 - J < \sigma < d/2 + 1 + \varepsilon.$$

For the proof see [23].

5 Non-linear twists

5.1 Meromorphic continuation

Let $F \in S_d^\#$, $d > 0$. For a real $\alpha > 0$ and $\sigma > 1$ we define the *non-linear twist* by the formula

$$F(s, \alpha) = \sum_{n=1}^{\infty} \frac{a(n)}{n^s} e(-n^{1/d}\alpha).$$

We write

$$n_\alpha = q_F \, d^{-d} \alpha^d$$

and

$$a(n_\alpha) = \begin{cases} a(n) & \text{if } n_\alpha \in \mathbb{N} \\ 0 & \text{otherwise.} \end{cases}$$

With this notation we have the following result.

Theorem 5.1.1 ([24]) *$F(s, \alpha)$ has meromorphic continuation to $\mathbb{C}$. Moreover, $F(s, \alpha)$ is entire if $a(n_\alpha) = 0$, while if $a(n_\alpha) \neq 0$ then $F(s, \alpha)$ has at most simple poles at the points*

$$s_k = \frac{d+1}{2d} - \frac{k}{d} - i\frac{\theta_F}{d}, \qquad k \geqslant 0,$$

with non-vanishing residue at s_0. Here θ_F is defined as in Section 3.3.

Proof Let $N > 2$ and $z_N = 1/N + 2\pi i\alpha$. For $\sigma \leqslant 1 + 1/d$ and sufficiently large positive c we have

$$F_N(s, \alpha) := \sum_{n=1}^{\infty} \frac{a(n)}{n^s} \exp\left(-n^{1/d} z_N\right)$$

$$= \frac{1}{2\pi i} \int_{c-i\infty}^{c+i\infty} F\left(s + \frac{w}{d}\right) \Gamma(w) z_N^{-w} \mathrm{d}w.$$

We shift the line of integration to the left and apply the functional equation to F. Next we move N to infinity. For $1 < \sigma < 1 + 1/d$ we obtain

$$F(s, \alpha) = R(s, \alpha) +$$

$$\omega Q^{1-2s} \sum_{n=1}^{\infty} \frac{\overline{a_F(n)}}{n^{1-s}} H_K\left(-\frac{i}{\widetilde{\beta}}\left(\frac{n}{n_\alpha}\right)^{1/d}, \widetilde{s}, \widetilde{\lambda}_1, \ldots, \widetilde{\lambda}_r, \widetilde{\mu}_1, \ldots, \widetilde{\mu}_r\right), \quad (5.1)$$

where

$$R(s, \alpha) = \operatorname*{Res}_{w=d(1-s)} F\left(s + \frac{w}{d}\right)\left(2\pi e^{i\pi/2}\alpha\right)^{-w} + \sum_{k=0}^{K} \frac{(-1)^k}{k!} F\left(s - \frac{k}{d}\right)(2\pi i\alpha)^k$$

is the corresponding sum of residues. Here $\widetilde{\beta}, \widetilde{s}, \ldots$ are defined as in Section 4.3.

We now consider the following two cases.

Case I: $n_\alpha \notin \mathbb{N}$. Since $R(s, \alpha)$ is entire, formula (5.1) together with Lemma 4.2.1 gives holomorphic continuation of $F(s, \alpha)$ to an entire function.

Case II: $n_\alpha \in \mathbb{N}$. Then (5.1) can be rewritten as follows:

$$F(s,\alpha) = \omega\, Q^{1-2s}\, \frac{\overline{a_F(n_\alpha)}}{n_\alpha^{1-s}}\, H_K\left(-i/\widetilde{\beta},\, \widetilde{s}\right) + E(s,\alpha)$$

with entire $E(s,\alpha)$. We now use Lemma 4.2.2 and the result follows.

Some extra arguments are needed to get uniformity. For this purpose we call *admissible* a family $\mathcal{F}$ of L-functions from $S^{\#}$ if:

(1) $d_F \ll 1$, $m_F \ll 1$,

(2) F has a γ-factor with $Q \gg 1$, $\lambda_j \gg 1$, $\mu_j \ll 1$,

(3) $\displaystyle\sum_{n \leqslant x} |a(n)| \ll x^{1+\varepsilon}$,

with implied constants depending on the family $\mathcal{F}$.

Examples Admissible families are:

(a) one element family $\mathcal{F} = \{F\}$ for a fixed $F \in S^{\#}$,

(b) Dirichlet L-functions associated with primitive characters,

(c) normalized L-functions associated with holomorphic modular forms of bounded weight,

(d) Dedekind zeta functions of number fields with bounded degrees.

Note however that the set $\{L(s + ik, \chi)\}_{k \in \mathbb{Z}}$ where χ is a fixed primitive Dirichlet character is not admissible. Also, the family of all Dedekind zeta functions is not admissible.

Let $\|x\|$ denote the distance of x from the nearest integer n such that $a_F(n) \neq 0$. Moreover, let

$$\delta(\alpha) = \begin{cases} \alpha & \text{if } a(n_\alpha) \neq 0 \\[2mm] \dfrac{\alpha}{\|\alpha\|} & \text{if } a(n_\alpha) = 0. \end{cases}$$

With this notation we have the following result.

Theorem 5.1.2 ([24]) *Let $\mathcal{F}$ be an admissible family, $\alpha > 0$, $\Delta \geqslant 2$ and $|s - s_k| \geqslant \dfrac{1}{4d_F}$. Then*

$$F(s,\alpha) \ll_{\mathcal{F},\Delta} q_F^{c-\sigma}(|t| + 2)^{c\Delta}\delta(\alpha)^{c\Delta}$$

uniformly for $F \in \mathcal{F}$, $-\Delta \leqslant \sigma \leqslant 2$, where $c = c(\mathcal{F})$. Moreover

$$\operatorname*{Res}_{s=s_k} F(s,\alpha) = \omega_F^* \, \overline{a(n_\alpha)}\, q_F^{\frac{1}{2}-s_k}\, n_\alpha^{s_k-1}\, c_k(\mathcal{F}) \qquad (k \geqslant 0)$$

where $c_k(\mathcal{F}) = O_{\mathcal{F},k}(1)$ uniformly for $F \in \mathcal{F}$ and $c_0(F) \neq 0$.

5.2 Some consequences

Let $\phi(u)$ be a smooth function on $(0, \infty)$ with compact support. We denote by $\widetilde{\phi}(s)$ its Mellin transform

$$\widetilde{\phi}(s) = \int_0^\infty \phi(u) u^{s-1} \mathrm{d}u.$$

Clearly $\widetilde{\phi}(s)$ is an entire function.

For $F \in S^{\#}$, $\alpha > 0$ and $x > 0$, we write

$$S_F(x) = \sum_{n=1}^\infty \phi\left(\frac{n}{x}\right) a(n) \, e(-n^{1/d}\alpha).$$

Then we have the following result.

Theorem 5.2.1 ([24]) *Let $\mathcal{F}$ be an admissible family and let $\alpha > 0$. Then for every $A > 0$ we have*

$$S_F(x) = \omega_F^* \, \frac{\overline{a(n_\alpha)}}{n_\alpha} \, q_F^{1/2} \times$$

$$\sum_{k \leqslant d_F A + (d_F+1)/2} c_k(F) \, \widetilde{\phi}(s_k) \left(\frac{x n_\alpha}{q_F}\right)^{s_k} + O_{\mathcal{F}, \phi, A}\left(q_F^{c+A} \delta(\alpha)^{cA} x^{-A}\right) \quad (5.2)$$

uniformly for $F \in \mathcal{F}$ and $n_\alpha \gg_{\mathcal{F}} 1$, where $c = c(\mathcal{F})$.

Proof The proof is standard. We write the inverse Mellin transform

$$S_F(x) = \frac{1}{2\pi i} \int_{2-i\infty}^{2+i\infty} F(s, \alpha) \widetilde{\phi}(s) x^s \mathrm{d}s$$

and shift the line of integration to the left. Since ϕ is smooth, for every positive A we have

$$\left|\widetilde{\phi}(\sigma + it)\right| \ll_A (|t| + 2)^{-A}$$

uniformly in every vertical strip $a \leqslant \sigma \leqslant b$. By Theorem 5.1.2, $F(s, \alpha)$ grows polynomially on vertical lines and hence the main term in (5.2) is the sum of residues.

Another application is as follows.

Theorem 5.2.2 ([24]) *Let $F \in S_d^{\#}$, $d \geqslant 1$. Then for every polynomial P we have*

$$\sum_{n \leqslant x} a(n) = x P(\log x) + \Omega\left(x^{(d-1)/(2d)}\right).$$

Proof For simplicity we consider in detail the case of an entire F. The case of non-entire F is more complicated but similar in principle.

We can assume $P \equiv 0$ since otherwise the result is trivial. Suppose that

$$\sum_{n \leqslant x} a(n) = o\big(x^{(d-1)/(2d)}\big).$$

Then by partial summation we have

$$\sum_{n > y} a(n)n^{-s} = o\big(y^{(d-1)/(2d)-\sigma}\big)$$

for $\sigma > (d+1)/(2d)$. For $\overline{\alpha} > 0$ with $a(n_{\overline{\alpha}}) \neq 0$ we have

$$F(s, \overline{\alpha}) - e(\overline{\alpha})F(s) = -\frac{2\pi i \,\overline{\alpha}}{d} \int_1^\infty \sum_{n > y} \frac{a(n)}{n^s} \, y^{1/d-1} e(-y^{1/d} \,\overline{\alpha}) \, dy$$

$$= o\left(\frac{1}{\sigma - (d+1)/(2d)}\right) \tag{5.3}$$

as σ tends to $(d+1)/(2d)$ from the right. By Theorem 5.1.1, $F(s, \overline{\alpha})$ has a pole at $s_0 = (d+1)/(2d) - i\theta_F/d$ and hence

$$F(\sigma - i\theta_F/d, \, \overline{\alpha}) - e(\overline{\alpha})F(\sigma - i\theta_F/d) \gg \frac{1}{\sigma - (d+1)/(2d)}$$

as $\sigma \to \big((d+1)/(2d)\big)^+$, a contradiction. The result therefore follows.

We remark that the standard way of proving results of this type is by the use of Voronoi-type identities. The advantage of the approach presented here is that it clearly shows a reason for the Ω-estimate, namely the pole of $F(s, \overline{\alpha})$. Note that the initial L-function $F(s)$ is regular for $s \neq 1$, and the special meaning of s_0 is not evident if one deals only with $F(s)$.

Let $\sigma_c(F)$ denote the abscissa of convergence of $F \in S^{\#}$.

Corollary 5.2.1 ([24]) *Let $F \in S_d^{\#}$, $d \geqslant 1$. Then the abscissa of convergence of F satisfies*

$$\sigma_c(F) \geqslant \frac{d-1}{2d}.$$

The result is trivial if $F(s)$ has a pole at $s = 1$. Otherwise, by a well known formula (cf. [44]) and by Theorem 5.2.2, we have

$$\sigma_c(F) = \limsup_{x \to \infty} \frac{\log\left|\sum_{n \leqslant x} a(n)\right|}{\log x} \geqslant \frac{d-1}{2d}.$$

6 Structure of the Selberg class: $d = 1$

6.1 The case of the extended Selberg class

Let $\Re \xi = \eta \in \{-1, 0\}$ and

$$\mathcal{X}(q,\xi) = \begin{cases} \{\chi \pmod q) : \chi(-1) = 1\} & \text{if } \eta = -1 \\ \{\chi \pmod q) : \chi(-1) = -1\} & \text{if } \eta = 0. \end{cases}$$

Moreover, for a Dirichlet character χ let

$$\omega_{\chi^*} = \tau(\chi^*)/(i^{a(\chi)} f_\chi)$$

denote the root number of the corresponding Dirichlet L-function. Here f_χ is the conductor of χ.

For a triplet (q, ξ, ω^*), where q is positive and ξ, ω^* are complex numbers with $|\omega^*| = 1$, we denote by $S_1^\#(q, \xi, \omega^*)$ the set of $F \in S_1^\#$ such that

$$q_F = q, \quad \xi_F = \xi, \quad \omega_F^* = \omega^*.$$

Then of course $S_1^\#$ is the disjoint union of these subclasses. Moreover, we write

$$V_1^\#(q, \xi, \omega^*) = S_1^\#(q, \xi, \omega^*) \cup \{0\}.$$

The structure of $S_1^\#$ is completely described by the following theorem.

Theorem 6.1.1 ([20]) *Let $F \in S_1^\#$. Then*

(1) *$q_F \in \mathbb{N}$ and $\eta_F = \Re\xi_F \in \{-1, 0\}$;*

(2) *the sequence $a(n)n^{i\theta_F}$ is periodic with period q_F;*

(3) *every $F \in S_1^\#(q, \xi, \omega^*)$, with $q \in \mathbb{N}$, $\eta = \Re\xi \in \{-1, 0\}$, $|\omega^*| = 1$ can be uniquely written as*

$$F(s) = \sum_{\chi \in \mathcal{X}(q,\xi)} P_\chi(s + i\theta_F) L(s + i\theta_F, \chi^*)$$

where $P_\chi \in S_0^\#(q/f_\chi, \omega^* \overline{\omega}_{\chi^*})$. Moreover, $P_{\chi_0}(1) = 0$ if $\theta \neq 0$.

(4) *$V_1^\#(q, \xi, \omega^*)$ is a real vector space and*

$$\dim_\mathbb{R} V_1^\#(q, \xi, \omega^*) = \begin{cases} [\frac{1}{2}q] & \text{if } \xi = -1 \\ [\frac{1}{2}(q - 1 - \eta)] & \text{otherwise.} \end{cases}$$

Proof For $d = 1$, the non-linear twist discussed in Section 5 becomes the additive linear twist:

$$F(s, \alpha) = \sum_{n=1}^\infty \frac{a(n)}{n^s} e(-n\alpha).$$

We have

$$F(s, \alpha) = F(s, \alpha + 1). \tag{6.1}$$

By Theorem 5.1.1 we know that this function has a pole at

$$s = 1 - i\theta_F \tag{6.2}$$

if $\alpha = n/q_F$ and $a(n) \neq 0$. Hence $\alpha q_F \in \mathbb{N}$ implies $(\alpha+1)q_F \in \mathbb{N}$ and therefore q_F is a positive integer. Moreover, the residue of $F(s, n/q_F)$ at (6.2) equals

$$\omega^* Q^{2i\theta_F - 1} \frac{\overline{a(n)}}{n^{i\theta_F}} \operatorname*{Res}_{s=1-i\theta_F} H_K(-i/\beta, s).$$

Using (6.1) we see that the same residue equals

$$\omega^* Q^{2i\theta_F - 1} \frac{\overline{a(n + q_F)}}{(n + q_F)^{i\theta_F}} \operatorname*{Res}_{s=1-i\theta_F} H_K(-i/\beta, s).$$

Since $\operatorname*{Res}_{s=1-i\theta_F} H_K(-i/\beta, s) \neq 0$, we see that the sequence

$$c(n) := a(n)n^{i\theta_F} \tag{6.3}$$

is periodic with period q_F.

For $\sigma > 1$ we have

$$F(s - i\theta_F) = \sum_{n=1}^{\infty} \frac{c(n)}{n^s}$$

$$= \sum_{\substack{d \mid q_F}} \sum_{\substack{n \geqslant 1 \\ (n, q_F) = d}} \frac{c(n)}{n^s}$$

$$= \sum_{\substack{d \mid q_F}} \frac{1}{d^s} \sum_{\substack{n \geqslant 1 \\ (n, q_F/d) = 1}} \frac{c(nd)}{n^s}.$$

We observe that for a fixed $d \mid q_F$, the function defined by

$$c_d(n) := \begin{cases} c(nd) & \text{if } (n, q_F/d) = 1 \\ 0 & \text{otherwise} \end{cases}$$

is periodic with period q_F/d. We may therefore write

$$c_d(n) = \sum_{\chi \pmod{q_F/d}} c_{d,\chi} \chi(n)$$

for certain complex numbers $c_{d,\chi}$. Hence

$$F(s - i\theta_F) = \sum_{d|q_F} \sum_{\chi \pmod{q_F/d}} \frac{c_{d,\chi}}{d^s} L(s, \chi)$$

$$= \sum_{d|q_F} \sum_{\chi \pmod{q_F/d}} \frac{c_{d,\chi}}{d^s} \prod_{p \mid \frac{q_F}{df_\chi}} \left(1 - \frac{\chi(p)}{p^s}\right) L(s, \chi^*)$$

$$= \sum_{\chi \pmod{q_F}} P_\chi(s) L(s, \chi^*),$$

where P_χ is a suitable Dirichlet polynomial.

Now we write the functional equations for $F(s - i\theta_F)$ and for each $L(s, \chi^*)$. They have to agree, and hence after long but rather standard computations we prove that the summation is over characters of the same parity, in fact over $\mathcal{X}(q_F, \xi_F)$. Moreover, we prove at the same time that the Dirichlet polynomials P_χ have to belong to $S_0^\#(q_F/f_\chi, \omega^* \overline{\omega}_{\chi^*})$. Since we have an explicit description of the functions from $S_1^\#(q, \xi, \omega^*)$, it is a matter of routine calculation to compute the dimension of $V_1^\#(q, \xi, \omega^*)$. Hence the theorem follows.

6.2 The case of the Selberg class

If $F \in S_1$ then the coefficients $a(n)$ are multiplicative. This gives a strong restriction on F.

Theorem 6.2.1 ([20]) *Let $F \in S_1$. If $q_F = 1$, then $F(s) = \zeta(s)$. If $q_F \geqslant 2$, then there exists a primitive Dirichlet character $\chi \pmod{q_F}$ with $\chi(-1) = -(2\eta_F + 1)$ such that $F(s) = L(s + i\theta_F, \chi)$.*

Proof Let n and m be two positive integers coprime to the conductor q_F. Using the Chinese Remainder Theorem we find $a \in \mathbb{N}$ such that $(m + aq_F, n) = 1$. Then, keeping the notation introduced in (6.3), we have

$$c(nm) = c((m + aq_F)n) = c(m + aq_F)c(n) = c(n)c(m).$$

Hence $c(n)$ is completely multiplicative on integers coprime to the conductor. Therefore

$$c(n) = \chi(n)$$

for a certain Dirichlet character $\chi \pmod{q_F}$ and for all integers n coprime to q_F. If $q_F = 1$ then $\theta_F = 0$ and we have $F(s) = \zeta(s)$. For $q_F > 1$ we have equality of local factors

$$F_p(s) = L_p(s + i\theta_F, \chi)$$

for all primes p not dividing q_F. Hence by the multiplicity one principle (Theorem 2.3.2) we have $F(s) = L(s + i\theta_F, \chi^*)$ and the result follows.

7 Structure of the Selberg class: $1 < d < 2$

7.1 Basic identity

Let $F \in S_d^{\#}$, $1 < d < 2$. For positive α and sufficiently large σ we write

$$F(s, \alpha) = \sum_{n=1}^{\infty} \frac{a(n)}{n^s} e(-n\alpha).$$

Moreover, let

$$\kappa = \frac{1}{d-1}, \qquad A = (d-1)q_F^{-\kappa},$$

$$s^* = \kappa\left(s + \frac{d}{2} - 1 + i\theta_F\right), \qquad \sigma^* = \Re s^*,$$

$$D(s, \alpha) = \sum_{n=1}^{\infty} \frac{\overline{a(n)}}{n^s} e\left(A\left(\frac{n}{\alpha}\right)^{\kappa}\right).$$

With this notation we have the following result.

Theorem 7.1.1 ([23]) *Let* $1 < d < 2$, $F \in S_d^{\#}$, $\alpha > 0$, *and let* $J \geqslant 1$ *be an integer. Then there exists a constant* $c_0 \neq 0$ *and polynomials* $P_j(s)$ *with* $0 \leqslant j \leqslant J - 1$ *and* $P_0(s) = c_0$ *identically, such that for* $\sigma^* > \sigma_a(F)$

$$F(s, \alpha) = q_F^{\kappa s} \sum_{j=0}^{J-1} \alpha^{\kappa(ds - d/2 + i\theta_F + j)} P_j(s) D(s^* + j\kappa, \alpha) + G_J(s, \alpha), \quad (7.1)$$

where $G_J(s, \alpha)$ *is holomorphic for* s *in the half-plane* $\sigma^* > \sigma_a(F) - \kappa J$ *and continuous for* $\alpha > 0$.

Proof Let $z_N := \dfrac{1}{N} + 2\pi i\alpha$, where N and α are real and positive, and let

$$F_N(s, \alpha) = \sum_{n=1}^{\infty} \frac{a(n)}{n^s} e^{-nz_N}.$$

Using the inverse Mellin transform and the functional equation for F we have

$$F_N(s, \alpha) = R_N(s, \alpha) + \chi_1(s)F(s) - \chi_2(s)z_N F(s - 1)$$

$$+ \omega Q^{1-2s} \sum_{n=1}^{\infty} \frac{\overline{a(n)}}{n^{1-s}} H_c\left(\frac{n}{Q^2 z_N}, s\right)$$

where $H_c(z, s)$ is the incomplete Fox hypergeometric function defined in Section 4.4,

$$\chi_1(s) = \begin{cases} 1 & \text{if } c > 0 \\ 0 & \text{if } c < 0, \end{cases} \qquad \chi_2(s) = \begin{cases} 1 & \text{if } c > 1 \\ 0 & \text{if } c < 1, \end{cases}$$

and

$$R_N(s, \alpha) = \operatorname*{Res}_{w=1-s} \left(F(s+w)\Gamma(w)z_N^{-w} \right).$$

Now we let $N \to \infty$ and use Lemma 4.4.3. After some computations we prove our theorem. For details see [23].

Theorem 7.1.1 has some interesting immediate consequences.

Corollary 7.1.1 ([23]) *Every $F \in S_d^{\#}$ with $1 < d < 2$ is entire.*

Indeed, for $1 < d < 2$ we have

$$\sigma^* = \Re s^* = \frac{\sigma + d/2 - 1}{d - 1}$$

and hence $\sigma^* > 1$ for $\sigma > d/2$. Hence the right hand side of (7.1) is holomorphic for $\sigma > d/2$. In particular, $F(s) = F(s, 1)$ is holomorphic at $s = 1$.

We can now give a further short proof of Theorem 2.2.3.

Corollary 7.1.2 ([23]) *We have $S_d^{\#} = \varnothing$ for $0 < d < 1$.*

Proof Let $F(s)$ be an L-function in $S_d^{\#}$, with $0 < d < 1$. We can assume without loss of generality that $F(1) \neq 0$. Indeed, when $F(1) = 0$ we replace $F(s)$ by an appropriate shift $F(s + i\theta)$. Then $F(s)\zeta(s)$ belongs to $S_d^{\#}$, $1 < d < 2$, and has a pole at $s = 1$, a contradiction with Corollary 7.1.1.

7.2 Fourier transform method

Let X be a sufficiently large integer, $\varepsilon > 0$, ν, ρ positive constants with $\rho > \nu + 1$, $\omega(y) \in C_0^{\infty}(\mathbb{R})$ with support contained in $[-\nu, \nu]$ and such that $0 \leqslant \omega(y) \leqslant 1$. Let $\widehat{\omega}(x)$ be the Fourier transform of ω, and

$$\sigma_{\varepsilon}^* = \sigma_a(F) - \kappa - \varepsilon, \qquad s_{\varepsilon} = (d-1)\sigma_{\varepsilon}^* - \frac{d}{2} + 1 - i\theta_F.$$

Moreover, let

$$2 \leqslant \omega_1 \leqslant \omega_2, \qquad X \leqslant x \leqslant 2X,$$

$$c_1 = e\left(\frac{1}{8}\right)2^{-1/2}, \qquad g(s, y) = \left(1 + \left(\frac{y}{A}\right)^{d-1}\right)^{\kappa(ds - d/2 + i\theta_F)},$$

$$\eta(y) = A\left(1 + \left(\frac{A}{y}\right)^{d-1}\right)^{-\kappa}, \qquad y_0 = y_0\left(\frac{n}{x}\right) = A\left(\left(\frac{n}{x}\right)^{1/d} - 1\right)^{\kappa},$$

$$f(x, n) = A(n^{1/d} - x^{1/d})^{\kappa d}, \qquad \lambda\left(\frac{n}{x}\right) = \overline{c_1}\left(\frac{x}{n}\right)^{\sigma_{\varepsilon}^* + \kappa/2} \frac{g(s_{\varepsilon}, y_0)\,\omega(y_0 - \rho)}{\sqrt{|\eta''(y_0)|}},$$

$$\Sigma_1(x) = \sum_{n=X}^{2X} \overline{a(n)} \left(\frac{x}{n}\right)^{\sigma_\varepsilon^*} \widehat{\omega}(x^\kappa - n^\kappa)\, e\big(\rho(n^\kappa - x^\kappa)\big), \tag{7.2}$$

$$\Sigma_2(x) = \sum_{\omega_1 X \leqslant n \leqslant \omega_2 X} \overline{a(n)}\, \lambda\left(\frac{n}{x}\right) e\big(f(x,n)\big).$$

Lemma 7.2.1 ([23]) *For every test function $\omega(y)$ there exists a shift ρ such that $2 \leqslant \omega_1 < \omega_2$ and, for every $\varepsilon > 0$,*

$$\Sigma_1(x) = x^{-\kappa/2}\Sigma_2(x) + O_\varepsilon\big(X^{\sigma_a(F)-\kappa+\varepsilon}\big)$$

as $X \to \infty$, uniformly for $X \leqslant x \leqslant 2X$.

Proof For $\sigma^* > \sigma_a(F)$ we have, using Theorem 7.1.1 with $J = 1$,

$$F(s,\alpha) = c_0 q^{\kappa s} \alpha^{\kappa(ds - d/2 + i\theta_F)} D(s^*, \alpha) + h(s),$$

where $h(s)$ is holomorphic for $\sigma > \sigma_a(F) - \kappa$. We use periodicity in α. We replace α in the above formula by $\alpha + 1$ and subtract both formulae. After a suitable change of variables we arrive at the following equality $(y > 0)$:

$$\sum_{n=1}^{\infty} \frac{\overline{a(n)}}{n^{s^*}}\, e(n^\kappa y) = g(s,y) \sum_{n=1}^{\infty} \frac{\overline{a(n)}}{n^{s^*}}\, e(n^\kappa \eta(y)) + h_1(s^*, y), \tag{7.3}$$

where $h_1(s^*, y)$ is holomorphic for $\sigma^* > \sigma_a(F) - \kappa$ and continuous for $y > 0$. Moreover,

$$g(s,y) = \left(1 + \left(\frac{y}{A}\right)^{d-1}\right)^{\kappa(ds - d/2 + i\theta_F)}$$

and

$$\eta(y) = A\left(1 + \left(\frac{A}{y}\right)^{d-1}\right)^{-\kappa}.$$

We now compute

$$I := \int_{-\infty}^{\infty} \omega(y - \rho) \left(\sum_{n=1}^{\infty} \frac{\overline{a(n)}}{n^{s^*}}\, e(n^\kappa y)\right) e(-x^\kappa y)\, dy \qquad (\rho > \nu + 1).$$

Integrating term by term we obtain

$$I = \sum_{n=1}^{\infty} \frac{\overline{a(n)}}{n^{s^*}}\, \widehat{\omega}(x^\kappa - n^\kappa)\, e\big(\rho(n^\kappa - x^\kappa)\big).$$

Using (7.3) we have

$$I = \sum_{n=1}^{\infty} \frac{\overline{a(n)}}{n^{s^*}} \int_{-\infty}^{\infty} g(s,y)\, \omega(y - \rho)\, e(n^\kappa \eta(y) - x^\kappa y)\, dy + E,$$

where E stands for a negligible error term. Now we apply the saddle point method to evaluate integrals and the result follows after some computations (see [23] for details).

Corollary 7.2.1 ([23]) $S_d^\# = \varnothing$ for $1 < d < \dfrac{3}{2}$.

Proof We apply Lemma 7.2.1. We take $x = n$, an integer of size X (X is positive and large). We have

$$\Sigma_1(n) = \overline{a(n)}\,\widehat{\omega}(0) + O_K(x^{-K})$$

for every positive K, by the fast decay of the Fourier transform and by $\kappa > 1$. Indeed, we have

$$(n+1)^\kappa - n^\kappa \asymp n^{\kappa-1} \gg X^{\kappa-1}$$

and hence

$$\widehat{\omega}(k^\kappa - n^\kappa) \ll_K X^{-K} \qquad (k \neq n).$$

Using Lemma 7.2.1 we therefore obtain

$$a(n) \ll X^{-\kappa/2} \sum_{n \sim X} |a(n)| + X^{\sigma_a(F)-\kappa} \ll X^{\sigma_a(F)-\kappa/2+\varepsilon}.$$

Consequently

$$\sum_{n=1}^{\infty} \frac{|a(n)|}{n^\sigma} < \infty$$

for $\sigma > 1 + \sigma_a(F) - \kappa/2$. If $d < 3/2$ then $\kappa > 2$, and we get a contradiction with the definition of the abscissa of absolute convergence. Hence the corollary follows.

7.3 Rankin-Selberg convolution

Definition $\qquad F \otimes \overline{F}(s) = \sum_{n \geq 1} |a(n)|^2 n^{-s} \qquad (\Re s > 2\sigma_a(F)).$

Lemma 7.3.1 *Let* $1 < d < 2$ *and* $F \in S_d^\#$. *Then* $F \otimes \overline{F}(s)$ *is holomorphic for* $\sigma > \sigma_a(F) - \kappa$ *apart from a simple pole at* $s = 1$.

Proof Our starting point is formula (7.3) which we rewrite for short as follows:

$$L(s^*, y) = R(s^*, y) + h_1(s^*, y). \tag{7.4}$$

Let $\theta(y)$ be a positive, C^∞ function with compact support contained in $(0, \infty)$. We compute the following convolution

$$\mathcal{F}(s^*) = \int_{-\infty}^{\infty} \theta(y) L(s^*, y) \overline{L}(s^*, y) \, \mathrm{d}y.$$

We change the order of integration and summation. Since $\kappa > 1$, only the diagonal terms matter here. Thus after integrating term by term we obtain

$$\mathcal{F}(s^*) = F \otimes \overline{F}(2s^*) \int_{-\infty}^{\infty} \theta(y) \, \mathrm{d}y + h(s^*),$$

where $h(s^*)$ is entire. Using (7.4) we obtain

$$\mathcal{F}(s^*) = F \otimes \overline{F}(2s^*) \int_{-\infty}^{\infty} \theta(y) h(y)^{2s^*-1} \, \mathrm{d}y + h_2(s^*),$$

where $h_2(s^*)$ is holomorphic for $\sigma^* > \sigma_a(F) - \kappa$. Writing $w = 2s^*$ we have

$$h_3(w) \, F \otimes \overline{F}(w) = h_4(w),$$

where

$$h_3(w) = \int_{-\infty}^{\infty} \theta(y)(1 - h(y))^{w-1} \, \mathrm{d}y$$

and $h_4(w)$ is holomorphic for $\Re w > \sigma_a(F) - \kappa$. Note that the entire function $h_3(w)$ has a simple zero at $w = 1$, while by a suitable choice of $\theta(y)$ we can ensure that $h_3(w) \neq 0$ for $w \in [\sigma_a(F) - \kappa, \ 2\sigma_a(F)] \setminus \{1\}$. Therefore the only possible pole of $F \otimes \overline{F}(w)$ on the half-plane $\Re w > \sigma_a(F) - \kappa$ is at $w = 1$. Moreover $w = 1$ is a simple pole of $F \otimes \overline{F}(w)$, as follows from Landau's theorem on Dirichlet series. The lemma is proved.

Corollary 7.3.1 ([23]) *For $x \to \infty$ we have*

$$\sum_{n \leqslant x} |a(n)|^2 \sim c_0 X$$

with a positive constant c_0.

7.4 Non existence of L-functions of degrees $1 < d < 5/3$

Theorem 7.4.1 ([23]) *We have $S_d^{\#} = \varnothing$ for $1 < d < \dfrac{5}{3}$.*

Proof We consider the integral

$$J(X) = \int_X^{2X} |\Sigma_1(x)|^2 \, e(x) \, \mathrm{d}x,$$

where $X > 0$ is sufficiently large and $\Sigma_1(x)$ is defined by (7.2). Moreover, let $\Delta = X^{1-\kappa+\delta}$ for a sufficiently small positive δ. Owing to the decay of $\widehat{\omega}(x)$ and since $\kappa > 1$ we have

$$J(X) = (1 + O(\Delta)) \sum_{n=X}^{2X} |a(n)|^2 \int_{n-\Delta}^{n+\Delta} \left(\frac{x}{n}\right)^{2\sigma_c^*} |\widehat{\omega}(x^\kappa - n^\kappa)|^2 \, \mathrm{d}x + O(X^{-M})$$

for every $M > 0$. It is easy to see that the integrals on the right hand side are $\gg X^{1-\kappa}$. Hence, using Corollary 7.3.1, we have

$$J(X) \gg X^{2-\kappa}. \tag{7.5}$$

In order to obtain the upper bound for $J(X)$ we apply Lemma 7.2.1, thus getting

$$J(X) \ll X^{-\kappa} \left| \sum_{\omega_1 X \leqslant n,m \leqslant \omega_2 X} \overline{a(n)}\, a(m)\, I_{n,m} \right| + X^{2\sigma_a(F)+1-2\kappa+\varepsilon}, \tag{7.6}$$

where

$$I_{n,m} = \int_X^{2X} \lambda\!\left(\frac{n}{x}\right) \overline{\lambda\!\left(\frac{m}{x}\right)}\, e\big(f(x,n) - f(x,m) + x\big)\, \mathrm{d}x.$$

If $|n - m| \not\asymp X^{2-\kappa}$ then

$$\left| \frac{\partial}{\partial x}\big(f(x,n) - f(x,m) + x\big) \right| \gg |n - m|\, X^{\kappa-2}.$$

Hence, by the first derivative test, we get

$$I_{n,m} \ll \frac{X^{2-\kappa}}{|n - m|}$$

for $0 \neq |n - m| \not\asymp X^{2-\kappa}$, and

$$I_{n,n} \ll 1.$$

For $|n - m| \asymp X^{2-\kappa}$ we have

$$\left| \frac{\partial^2}{\partial x^2}\big(f(x,n) - f(x,m) + x\big) \right| \gg 1/X,$$

whence by the second derivative test we obtain

$$I_{n,m} \ll X^{1/2}.$$

Inserting these estimates into (7.6) we obtain after some calculations

$$J(X) \ll X^{7/2-\kappa}. \tag{7.7}$$

Comparing (7.5) and (7.7) we obtain $\kappa \leqslant 3/2$, i.e. $d \geqslant 5/3$, and the result follows.

7.5 *Dulcis in fundo*

We end these lecture notes with the following converse theorems concerning the Riemann zeta function and the Dirichlet L-functions. Formulations are simple but proofs, although short, heavily depend on the main results of Chapters 5, 6 and 7, and therefore are rather deep.

Theorem 7.5.1 ([24]) *Let $F \in S_d$ with $d \geqslant 1$. If the series*

$$\sum_{n=1}^{\infty} \frac{a_F(n) - 1}{n^s}$$

converges for $\sigma > 1/5 - \delta$ with some $\delta > 0$, then $F(s) = \zeta(s)$.

Theorem 7.5.2 ([24]) *Let $F \in S_d$ with $d \geqslant 1$. If the series*

$$\sum_{n=1}^{\infty} \frac{a_F(n)}{n^s}$$

converges for $\sigma > 1/5 - \delta$ with some $\delta > 0$, then $F(s) = L(s + i\theta, \chi)$ with some $\theta \in \mathbb{R}$ and a primitive Dirichlet character $\chi \, (mod \, q)$, $q > 1$.

Proof We prove Theorem 7.5.2. The proof of Theorem 7.5.1 is similar but needs some modifications (cf. [24]).

From Corollary 5.2.1 we know that the abscissa of convergence of $F(s)$ is $\geqslant (d-1)/(2d)$. Therefore, according to our convergence assumption, $d < 5/3$. Using Theorem 7.4.1 we conclude that $d = 1$, and the result follows by an application of Theorem 6.2.1.

References

[1]   J. Arthur, *Automorphic representations and number theory*, Canad. Math. Soc. Conf. Proc., Vol. 1, 1981, 3-54.

[2]   S. Bochner, *On Riemann's functional equation with multiple gamma factors*, Ann. of Math. (2) 67 (1958), 29-41.

[3]   H. Bohr, *Über fastperiodische ebene Bewegungen*, Comment. Math. Helv. 4 (1932), 51-64.

[4]   A. Borel, *Automorphic L-functions*, Proc. Sympos. Pure Math., Vol. 33, Part 2, Amer. Math. Soc., Providence, R.I., 1979, 27-61.

[5]   B. L. J. Braaksma, *Asymptotic expansions and analytic continuations for a class of Barnes-integrals*, Compositio Math. 15 (1963), 239-431.

[6]   D. Bump, *Authomorphic forms and representations*, Cambridge University Press, 1997.

[7]   E. Carletti, G. Monti Bragadin, A. Perelli, *On general L-functions*, Acta Arith. 66 (1994), 147-179.

[8]   K. Chandrasekharan, R. Narasimhan, *Functional equations with multiple gamma factors and the average order of arithmetical functions*, Ann. of Math. 76 (1962), 93-136.

[9]   J. B. Conrey, A. Ghosh, *On the Selberg class of Dirichlet series: small degrees*, Duke Math. J. 72 (1993), 673-693.

[10] H. Davenport, *Multiplicative number theory*, 2nd ed., Springer-Verlag, Berlin-Heidelberg, 1980.

[11] A. Erdélyi, W. Magnus, F. Oberhettinger, F. G. Tricomi, *Higher Transcendental Functions*, Vol. I, McGraw-Hill, New York, 1953.

[12] S. Gelbart, *An elementary introduction to the Langlands program*, Bull. Amer. Math. Soc. 10 (1984), 177-219.

[13] R. Godement, H. Jacquet, *Zeta-functions of simple algebras*, Lecture Notes in Math., Vol. 260, Springer-Verlag, Berlin-Heidelberg, 1972.

[14] H. Heilbronn, *Zeta-functions and L-functions*, in: *Algebraic number theory*, J. W. S. Cassels, A. Fröhlich (eds.), Academic Press, London, 1967.

[15] A. Ivić, *The Riemann zeta-function*, Wiley, New York, 1985.

[16] H. Iwaniec, *Topics in classical automorphic forms*, Graduate Studies in Math., Vol. 17, Amer. Math. Soc., Providence, R.I., 1997.

[17] H. Jacquet, R. P. Langlands, *Automorphic forms on* GL(2), Lecture Notes in Math., Vol. 114, Springer-Verlag, Berlin-Heidelberg, 1970.

[18] H. Jacquet, J. A. Shalika, *On Euler products and the classification of automorphic representations, I*, Amer. J. Math. 103 (1981), 499-558.

[19] H. Jacquet, J. A. Shalika, *On Euler products and the classification of automorphic representations, II*, Amer. J. Math. 103 (1981), 777-815.

[20] J. Kaczorowski, A. Perelli, *On the structure of the Selberg class, I: $0 \leqslant d \leqslant 1$*, Acta Math. 182 (1999), 207-241.

[21] J. Kaczorowski, A. Perelli, *On the structure of the Selberg class, II: invariants and conjectures*, J. Reine Angew. Math. 524 (2000), 73-96.

[22] J. Kaczorowski, A. Perelli, *On the structure of the Selberg class, IV: basic invariants*, Acta Arith. 104 (2002), 97-116.

[23] J. Kaczorowski, A. Perelli, *On the structure of the Selberg class, V: $1 < d < 5/3$*, Invent. Math. 150 (2002), 485-516.

[24] J. Kaczorowski, A. Perelli, *On the structure of the Selberg class, VI: non-linear twists*, Acta Arith. 116 (2005), 315-341.

[25] J. Kaczorowski, A. Perelli, *Strong multiplicity one for the Selberg class*, C. R. Acad. Sci. Paris Sér. I Math. 332 (2001), 963-968.

[26] J. Kaczorowski, A. Perelli, *Factorization in the extended Selberg class*, Funct. Approx. Comment. Math. 31 (2003), 109-117.

[27] W. C. Winnie Li, *Number Theory with Applications*, Series on University Math., Vol. 7, World Scientific Publ. Co., River Edge, N.J., 1996.

[28] J. Martinet, *Character theory and Artin L-functions*, in: *Algebraic number fields; L-functions and Galois properties*, Proc. Sympos. Univ. Durham, 1975, A. Fröhlich (ed.), Academic Press, London-New York, 1977.

[29] T. Miyake, *Modular forms*, Springer-Verlag, Berlin-Heidelberg, 1989.

[30] M. Ram Murty, *Selberg's conjectures and Artin L-functions*, Bull. Amer. Math. Soc. 31 (1994), 1-14.

[31] M. Ram Murty, V. Kumar Murty, *Strong multiplicity one for Selberg's class*, C. R. Acad. Sci. Paris Sér. I Math. 319 (1994), 315-320.

[32] M. Ram Murty, *Stronger multiplicity one for Selberg's class*, in: *Harmonic analysis and number theory*, S. W. Drury, M. Ram Murty (eds.), CMS Conf. Proc., Vol. 21, Amer. Math. Soc., Providence, R.I., 1997, 133-142.

[33] W. Narkiewicz, *Elementary and analytic theory of algebraic numbers*, PWN, Warszawa, Springer-Verlag, Berlin-Heidelberg, 1990.

[34] A. Perelli, *General L-functions*, Ann. Mat. Pura Appl. (4) 130 (1982), 287-306.

[35] I. Piatetski-Shapiro, *Multiplicity one theorems*, in: *Automorphic forms, representations and L-functions*, A. Borel, W. Casselman (eds.), Proc. Sympos. Pure Math., Vol. 33, Part 1, Amer. Math. Soc., Providence, R.I., 1979, 209-212.

[36] K. Prachar, *Primzahlverteilung*, Springer-Verlag, Berlin-Heidelberg, 1978.

[37] H.-E. Richert, *Über Dirichletreihen mit Funktionalgleichung*, Publ. Inst. Math. Acad. Serbe Sci. 1 (1957), 73-124.

[38] A. Selberg, *Old and new conjectures and results about a class of Dirichlet series*, Proc. Amalfi Conference on Analytic Number Theory (Maiori, 1989), E. Bombieri et al. (eds.), Università di Salerno, 1992, 367-385; Collected papers, Vol. II, Springer-Verlag, Berlin-Heidelberg, 1991, 47-63.

[39] J.-P. Serre, *Représentations linéaires des groupes finis*, Hermann, Paris, 1971.

[40] F. Shahidi, *On non-vanishing of L-functions*, Bull. Amer. Math. Soc. 2 (1980), 462-464.

[41] K. Soundararajan, *Strong multiplicity one for the Selberg class*, Canad. Math. Bull. 47 (2004), 468-474.

[42] J. Tate, *Fourier analysis in number fields and Hecke's zeta-functions*, Thesis, Princeton, 1950; reproduced in: *Algebraic number theory*, J. W. S. Cassels, A. Fröhlich (eds.), Academic Press, London, 1967.

[43] E. C. Titchmarsh, *The theory of the Riemann zeta function*, 2nd ed., Clarendon Press, Oxford, 1988.

[44] E. C. Titchmarsh, *The theory of functions*, 2nd ed., Oxford University Press, 1939.

List of Participants

1. Adhikari Sukumar Das
   Harish-Chandra R.I., India
   adhikari@mri.ernet.in

2. Aliev Iskander
   Polish Academy of Science
   iskander@impan.gov.pl

3. Avanzi Roberto
   Univ. Essen
   mocenigo@exp-math.
   uni-essen.de

4. Basile Carmen Laura
   Imperial College, London
   laura.basile@ic.ac.uk

5. Bourqui David
   Univ. Grenoble 1
   bourqui@ujf-grenoble.fr

6. Broberg Niklas
   Chalmers Univ.
   nibro@math.chalmers.se

7. Browning Timothy
   Math. Inst., Oxford
   browning@maths.ox.ac.uk

8. Brüdern Jörg
   Univ. Stuttgart
   bruedern@mathematik.
   uni-stuttgart.de

9. Chamizo Fernando
   Univ. Autonoma Madrid
   fernando.chamizo@uam.es

10. Chiera Francesco
    Univ. di Padova
    chiera@math.unipd.it

11. Chinta Gautam
    Brown Univ.
    chinta@math.brown.edu

12. Corvaja Pietro
    Univ. di Udine
    corvaja@dimi.uniud.it

13. Dahari Arle Samuel
    Bar Ilan Univ.
    dahari@math.biu.ac.il

14. De Roton Anne
    Univ. de Bordeaux
    deroton@math.u-bordeaux.fr

15. Dvornicich Roberto
    Univ. di Pisa
    dvornic@dm.unipi.it

16. Elsholtz Christian
    Tech. Univ. Clausthal
    elsholtz@math.tu-clausthal.de

17. Esposito Francesco
    Univ. di Roma 1
    esposito@mat.uniroma1.it

18. Fischler Stéphane
    Ec. Normale Sup., Paris
    fischler@math.jussieu.fr

19. Friedlander John B.
    Univ. of Toronto
    frdlndr@math.toronto.edu
    (lecturer)

20. Garaev Moubariz
    Ac. Sinica, Taiwan
    garaev@math.sinica.edu.tw

21. Heath-Brown D. R.
    Math. Inst., Oxford
    rhb@maths.ox.ac.uk
    (lecturer)

22. Ivic Alexandar
    Serbian Acad. Sc., Beograd
    aivic@rgf.bg.ac.yu

23. Iwaniec Henryk
    Rutgers Univ.
    iwaniec@math.rutgers.edu
    (lecturer)

24. Kaczorowski Jerzy
    Adam Mickiewicz Univ., Poznań
    kjerzy@amu.edu.pl
    (lecturer)
25. Kadiri Habiba
    Univ. de Lille
    kadiri@agat.univ-lille1.fr
26. Kawada Koichi
    Iwate Univ.
    kawada@iwate-u.ac.jp
27. Khemira Samy
    Univ. Paris 6
    khemira@math.jussieu.fr
28. Languasco Alessandro
    Univ. di Padova
    languasc@math.unipd.it
29. Laporta Maurizio
    Univ. di Napoli
    laporta@matna2.dma.unina.it
30. Lau Yuk-Kam
    Hong Kong Univ.
    yklau@maths.hku.hk
31. Longo Matteo
    Univ. di Padova
    mlongo@math.unipd.it
32. Makatchev Maxim
    Univ. of Pittsburg
    maxim@pitt.edu
33. Marcovecchio Raffaele
    Univ. di Pisa
    marcovec@mail.dm.unipi.it
34. Marmi Stefano
    Univ. di Udine e SNS
    marmi@sns.it
35. Melfi Giuseppe
    Univ. Neuchatel
    Giuseppe.Melfi@unine.ch
36. Mitiaguine Anton
    Moscow State Univ.
    mityagin@dnttm.ru
37. Molteni Giuseppe
    Univ. di Milano
    giuseppe.molteni@mat.unimi.it
38. Ng Nathan
    Univ. Montreal
    nathan@dms.umontreal.ca
39. Obukhovski Andrey
    Voronezh St. Univ., Russia
    andrei@ob.vsu.ru
40. Pappalardi Francesco
    Univ. Roma Tre
    pappa@mat.uniroma3.it
41. Perelli Alberto
    Univ. di Genova
    perelli@dima.unige.it
    (editor)

42. Rocadas Luis
    UTAD, Portugal
    rocadas@utad.pt
43. Rodionova Irina
    Voronezh St. Univ., Russia
    rodirina@hotmail.com
44. Schlickewei Hans Peter
    Univ. Marburg
    hps@mathematik.uni-marburg.de
45. Skogman Howard
    SUNY Brockport
    hskogman@brockport.edu
46. Summerer Leonard
    ETH Zurich
    summerer@math.ethz.ch
47. Surroca Andrea
    Univ. Paris 6
    surroca@math.jussieu.fr
48. Tchanga Maris
    Steklov Inst., Moscow
    maris_changa@mail.ru
49. Traupe Martin
    mamt@math.tu-clausthal.de
50. Tubbs Robert
    Univ. of Colorado, Boulder
    tubbs@euclid.colorado.edu
51. Ubis Adrian
    Univ. Autonoma Madrid
    adrian.ubis@uam.es
52. Viola Carlo
    Univ. di Pisa
    viola@dm.unipi.it
    (editor)
53. Viviani Filippo
    Univ. di Roma 2
    viviani@mat.uniroma2.it
54. Vorotnikov Dmitry
    Voronezh St. Univ., Russia
    mitvorot@mail.ru
55. Welter Michael
    Univ. Köln
    mwelter@mi.uni-koeln.de
56. Zaccagnini Alessandro
    Univ. di Parma
    zaccagni@math.unipr.it
57. Zannier Umberto
    SNS, Pisa
    u.zannier@sns.it
58. Zhang Qiao
    Columbia Univ., NY, USA
    qzhang@math.columbia.edu
59. Zudilin Wadim
    Lomonosov St. Univ., Moscow
    wadim@ips.ras.ru

LIST OF C.I.M.E. SEMINARS

Published by C.I.M.E

Published by Ed. Cremonese, Firenze

Published by Ed. Liguori, Napoli

Published by Ed. Liguori, Napoli & Birkhäuser